BASIC CONCEPTS IN
BIOCHEMISTRY

BASIC CONCEPTS IN BIOCHEMISTRY

A STUDENT'S SURVIVAL GUIDE

HIRAM F. GILBERT, Ph.D.

Professor of Biochemistry
Baylor College of Medicine
Houston, Texas

McGraw-Hill, Inc.
Health Professions Division

New York   St. Louis   San Francisco
Auckland   Bogota   Caracus   Lisbon   London   Madrid
Mexico   Milan   Montreal   New Delhi   Paris   San Juan
Singapore   Sydney   Tokyo   Toronto

BASIC CONCEPTS IN BIOCHEMISTRY:
A STUDENT'S SURVIVAL GUIDE

1 2 3 4 5 6 7 8 9 0 DOC DOC 9 8 7 6 5 4 3 2 1

This book was set in Times Roman by Better Graphics, Inc. The editors were Gail Gavert and Lester A. Sheinis. The production supervisor was Clare Stanley. The text and cover were designed by Marsha Cohen.

R. R. Donnelley & Sons Company was printer and binder.

ISBN 0-07-023449-3

Library of Congress Cataloging-in-Publication Data

Gilbert, Hiram F.
    Basic concepts in biochemistry : a student's survival guide / Hiram F. Gilbert.
        p.   cm.
    Includes index.
    ISBN 0-07-023449-3
    1. Biochemistry.   I. Title.
    [DNLM: 1. Biochemistry.   QU 34 G464b]
QP514.2.G56   1991
612'.015—dc20
DNLM/DLC
for Library of Congress   91-19647
                                                                CIP

· C O N T E N T S ·

CHAPTER 15   INTEGRATION OF ENERGY METABOLISM   159

CHAPTER 16   UREA CYCLE   196

CHAPTER 17   PURINE METABOLISM   198

CHAPTER 18   PYRIMIDINE METABOLISM   202

CHAPTER 19   ONE-CARBON METABOLISM   205

CHAPTER 20   TRACKING CARBONS   208

CHAPTER 21   pH, pK_a, pRoblems   213

CHAPTER 22   THERMODYNAMICS
AND KINETICS   233

· P R O L O G U E ·

Basic Concepts in Biochemistry: A Student's Survival Guide is not a conventional book: It is not a review book or a textbook or a problem book. It is a book that offers help in two different ways—help in understanding the concepts of biochemistry and help in organizing your attack on the subject and minimizing the subject's attack on you.

This book presents what are often viewed as the more difficult concepts in an introductory biochemistry course and describes them in enough detail and in simple enough language to make them understandable. We surveyed first- and second-year medical students at a national student meeting asking them to list, in order, the parts of biochemistry they found most difficult to understand. The winner (or loser), by far, was integration of metabolism. Metabolic control, pH, and enzyme kinetics ran closely behind, with notable mention being given to molecular biology and proteins.

Biochemistry texts and biochemistry professors are burdened with the task of presenting facts, and the enormity of this task can get in the way of explaining concepts. Since I don't feel burdened by that necessity, I've only outlined most of the facts and concentrated on concepts. My rationale is that concepts are considerably easier to remember than facts and that concepts, if appropriately mastered, can minimize the amount of material that has to be memorized—you can just figure everything out when required. In *Basic Concepts in Biochemistry*, central concepts are developed in a stepwise fashion. The simplest concepts provide a review of what might have been forgotten, and the more complex concepts present what might not have been realized.

BASIC CONCEPTS IN
BIOCHEMISTRY

WHERE TO START

·

Instructions

What Do I Need to Know?

Instructions for Use

Studying and Exams

Trivia Sorter

· · · · · · · · · · ·

INSTRUCTIONS

Read for understanding. Read only what you don't know. Organize, organize, organize.

The first page of each chapter presents an index. A title-summary box for each section presents a short summary and memory jogger intended to be helpful for review. If you already know what the boxed terms mean and feel comfortable with them, don't bother to read the text section that follows—proceed until you find a heading you don't understand, and then read till you understand. The first rule (it may not really be the first rule, but it is a rule) is not to waste time reading things you already know.

Keep on not reading the text until you find something you don't understand—then read the text till you do. The sections are generally arranged in order of increasing complexity and build on previous sections. So if you screwed up and jumped in over your head, back up a section or two. Another option is just to look at the pictures. Pictures and diagrams,

if extensively annnotated and carefully designed (by you), can be an
enormous review aid.

WHAT DO I NEED TO KNOW?

You need to know only the things you will need later.

Medicine and biology are becoming increasingly molecular in nature,
so one answer to the question is that you need to know things down to the
last atom. *Everything* is not the right answer. You can't possibly learn it
all. Therefore, you will have to be selective.

Another answer is that you just need to know the things on the exam.
Later ends at the final. In reality, later may be longer than this. Try to
pick out the major concepts of biochemistry as you go along. Concepts
are generally easier to remember than factual details—particularly if the
concepts make sense.

INSTRUCTIONS FOR USE

Understand the concepts first. Make notes. Never use a colored
highlighter.

General concepts don't need to be memorized. Once you understand
them, they provide a framework to hang the rest of the material on. Since
they don't need to be memorized, they can be learned (or thought about)
almost anywhere. To remember something, write it down. Don't just
highlight it with a colored pen or pencil. Highlighting is a great way to
forget to read the material.

STUDYING AND EXAMS

Organize, understand, condense, memorize.

**• 1. ALWAYS REMEMBER THAT IT IS POSSIBLE TO BE A WORTH-
WHILE HUMAN BEING REGARDLESS OF (OR IN SPITE OF) HOW
MUCH BIOCHEMISTRY YOU KNOW.** This won't necessarily help
you with biochemistry, but it may help you keep your sanity.

• **2. MINIMIZE THE AMOUNT OF MATERIAL THAT YOU HAVE TO MEMORIZE.**   If you understand a general concept, you can often figure out the specific details rather than memorize them. For example, does phosphorylation activate or inactivate acetyl-CoA carboxylase? You could just memorize that it inactivates the enzyme. However, this wouldn't help when it came to the phosphorylation of glycogen synthase. Try the following line of reasoning. We store energy after eating and retrieve it between meals. Storage and retrieval of energy do not happen at the same time. Protein phosphorylation generally increases when you're hungry. Since both acetyl-CoA carboxylase and glycogen synthase are involved in energy storage (fat and glucose, respectively), they will both be inactivated by phosphorylation. For just two enzymes, it might be easier to just memorize all the regulatory behaviors—but for several hundred?

• **3. ARRANGE NOTES AND STUDY TIME IN ORDER OF DECREAS-ING IMPORTANCE.**   During the first (or even second and third) pass, you can't possibly learn everything biochemistry has to offer. Be selective. Learn the important (and general) things first. If you have enough gray matter and time, then pack in the details. Organize your notes the same way. For each topic (corresponding to about a chapter in most texts) write down a *short* summary of the really important concepts (no more than one to two pages). Don't write down the things that you already know, just the things you're likely to forget. Be really cryptic to save space, and use lots of diagrams. These don't have to be publication-quality diagrams; they only have to have meaning for you. The idea is to minimize the sheer volume of paper. You can't find yourself at finals time with a yellow-highlighted 1000-page text to review 2 days before the exam. An enormous amount of information can be crammed onto a diagram, and you learn a significant amount by creating diagrams. Use them extensively.

• **4. SORT OUT THE TRIVIA AND FORGET ABOUT IT.**   The most difficult part may be deciding what the important things actually are. After all, if you've never had biochemistry, it all sounds important (or none of it does). Use the trivia sorter below (or one of your own invention) to help with these decisions. To use this sorter, you must first set your trivia level. Your trivia level will depend on whether you just want to pass or want to excel, whether you want to devote a lot of time or a whole lot of time to biochemistry, and your prior experience. Once you set this level, make sure you know *almost everything* above this level and ignore almost everything below it. Setting your trivia level is not irreversible; the setting can be moved at any time. You should consider levels 7 to 10 as the minimal acceptable trivia level (passing). The trivia sorter shown here is generic. You can make your own depending on the exact demands of the course you're taking. Levels 21 and 22 might be too trivial for anybody to spend time learning (again, this is opinion).

TRIVIA SORTER

1. Purpose of a pathway—what's the overall function?
2. Names of molecules going into and coming out of the pathway
3. How the pathway fits in with other pathways
4. General metabolic conditions under which the pathway is stimulated or inhibited
5. Identity (by name) of control points—which steps of the pathway are regulated?
6. Identity (by name) of general regulatory molecules and the direction in which they push the metabolic pathway
7. Names of reactants and products for each regulated enzyme and each enzyme making or using ATP equivalents
8. Names of molecules in the pathway and how they're connected
9. Structural features that are important for the function of specific molecules in the pathway (this includes DNA and proteins)
10. Techniques in biochemistry, the way they work, and what they tell you
11. Molecular basis for the interactions between molecules
12. Genetic diseases and/or specific drugs that affect the pathway
13. Essential vitamins and cofactors involved in the pathway
14. pH
15. Enzyme kinetics
16. Specific molecules that inhibit or activate specific enzymes
17. Names of individual reactants and products for nonregulated steps
18. Chemical structures (ability to recognize, not draw)
19. Structures of individual reactants and products for all enzymes in pathway
20. Reaction mechanism (chemistry) for a specific enzyme
21. Cleavage specificity for proteases or restriction endonucleases
22. Molecular weights and quaternary structures

• **5. DON'T WASTE TIME ON ABSOLUTE TRIVIA UNLESS YOU HAVE THE TIME TO WASTE.**   It is possible to decide that something is just not worth remembering; for example, cleavage specificities of proteases or restriction endonucleases, and protein molecular weights, are obvious choices. You can set the "too trivial to bear" level anywhere you want. You could decide that glycolysis is just not worth knowing. However, if you set your limits totally in the wrong place, you will get another chance to figure this out when you repeat the course. The trivia line is an important line to draw, so think about your specific situation and the requirements of the course before you draw it.

PROTEIN STRUCTURE

·

·   ·   ·   ·   ·   ·   ·   ·   ·   ·   ·   ·

Proteins start out life as a bunch of amino acids linked together in a head-to-tail fashion—the primary sequence. The one-dimensional information contained in the primary amino acid sequence of cellular proteins is enough to guide a protein into its three-dimensional structure, to determine its specificity for interaction with other molecules, to determine its ability to function as an enzyme, and to set its stability and lifetime.

AMINO ACID STRUCTURE

Remember a few of the amino acids by functional groups. The rest are hydrophobic.

Remembering something about the structures of the amino acids is just one of those basic language things that must be dealt with since it crops up over and over again—not only in protein structure but later in metabolism. You need to get to the point that when you see Asp you don't think snake but see a negative charge. Don't memorize the amino acids down to the last atom, and don't spend too much time worrying about whether glycine is polar or nonpolar. Methylene groups ($-CH_2-$) may be important, but keeping track of them on an individual basis is just too much to ask. Organize the amino acids based on the functional group of the side chain. Having an idea about functional groups of amino acids will also help when you get to the biosynthesis and catabolism of amino acids. Might as well bite the bullet early.

HYDROPHILIC (POLAR)

• **CHARGED POLAR** *Acidic* ($-COO^-$) and *basic* ($-NH_3^+$) amino acid side chains have a charge at neutral pH and strongly "prefer" to be on the exterior, exposed to water, rather than in the interior of the protein. The terms *acidic* and *basic* for residues may seem a little strange. Asp and Glu are called acidic amino acids, although at neutral pH in most proteins,

FUNCTIONAL GROUP			AMINO ACID
Hydrophilic, Polar			
Acidic	Carboxylates	$-COO^-$	Asp, Glu
Basic	Amines	$-NH_3^+$	Lys, Arg, His
Neutral	Amides	$-CONH_2$	Asn, Gln
	Alcohols	$-OH$	Ser, Thr, Tyr
	Thiol	$-SH$	Cys
Hydrophobic, Apolar			
Aliphatic		$-CH_2-$	Ala, Val, Leu, Ile, Met
Aromatic		C Rings	Phe, Trp, Tyr
Whatever			Pro, Gly

Asp and Glu are not present in the acidic form (–COOH) but are present in the basic form (–COO⁻). So the acidic amino acids, Asp and Glu, are really bases (proton acceptors). The reason that Asp and Glu are called acidic residues is that they are such strong acids (proton donors) they have already lost their protons. Lys, Arg, and His are considered basic amino acids, even though they have a proton at neutral pH. The same argument applies: Lys, Arg, and His are such good bases (proton acceptors) that they have already picked up a proton at neutral pH.

Charged groups are usually found on the surface of proteins. It is very difficult to remove a charged residue from the surface of a protein and place it in the hydrophobic interior, where the dielectric constant is low. On the surface of the protein, a charged residue can be solvated by water, and it is easy to separate oppositely charged ions because of the high dielectric constant of water.[1] If a charged group is found in the interior of the protein, it is usually paired with a residue of the opposite charge. This is termed a *salt bridge*.

• **NEUTRAL POLAR**   These side chains are uncharged, but they have groups (–OH, –SH, NH, C=O) that can hydrogen-bond to water. In an unfolded protein, these residues are hydrogen-bonded to water. They prefer to be exposed to water, but if they are found in the protein interior they are hydrogen-bonded to other polar groups.

HYDROPHOBIC (APOLAR)

Hydrocarbons (both aromatic and aliphatic) do not have many (or any) groups that can participate in the hydrogen-bonding network of water. They're greasy and prefer to be on the interior of proteins (away from water). Note that a couple of the aromatics, Tyr and Trp, have O and N, and Met has an S, but these amino acids are still pretty hydrophobic. The hydrophobic nature usually dominates; however, the O, N, and S atoms often participate in hydrogen bonds in the interior of the protein.

---

[1] The dielectric constant is a fundamental and obscure property of matter that puts a number on how hard it is to separate charged particles or groups when they're in this material. In water, charge is easy to separate (water has a high dielectric constant). The charge distribution on water is uneven. It has a more positive end (H) and a more negative end (O) that can surround the charged group and align to balance the charge of an ion in water. This dipolar nature of water makes it easy for it to dissolve ionic material. Organic solvents like benzene or octane have a low dielectric constant and a more uniform distribution of electrons. They do not have polar regions to interact with ions. In these types of solvents, just as in the interior of a protein, it is very difficult to separate two oppositely charged residues.

INTERACTIONS

A few basic interactions are responsible for holding proteins to-gether. The properties of water are intimately involved in these interactions.

WATER

Water's important. Polar amino acid side chains can participate in hydrogen bonding to water, or hydrophobic side chains can interfere with it.

The properties of water dominate the way we think about the interactions of biological molecules. That's why many texts start with a lengthy, but boring, discussion of water structure, and that's why you probably do need to read it.

Basically, water is a polar molecule. The H—O bond is polarized—the H end is more positive than the O end. This polarity is reinforced by the other H—O bond. Because of the polarity difference, water is both a hydrogen-bond donor and a hydrogen-bond acceptor. The two hydrogens can each enter into hydrogen bonds with an appropriate acceptor, and the two lone pairs of electrons on oxygen can act as hydrogen-bond acceptors. Because of the multiple hydrogen-bond donor and acceptor sites, water interacts with itself. Water does two important things: It squeezes out oily stuff because the oily stuff interferes with the interaction of water with itself, and it interacts favorably with anything that can enter into its hydrogen-bonding network.

HYDROPHOBIC INTERACTION

Proteins fold in order to put as much of the greasy stuff out of contact with water as possible. This provides much of the "driving force" for protein folding, protein–protein interactions, and pro-tein–ligand interactions (Fig. 2-1).

The *driving force* for a chemical reaction is what makes it happen. It's the interaction that contributes the most to the decrease in free energy. For protein (and DNA) folding, it's the hydrophobic interaction

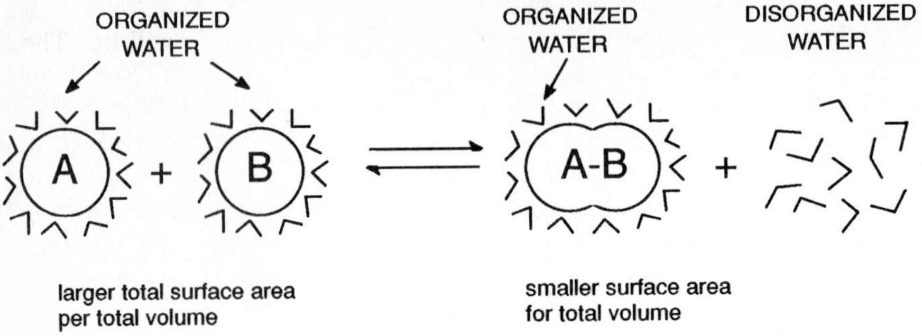

Figure 2-1   The Hydrophobic Interaction
As hydrophobic surfaces contact each other, the ordered water molecules that occupied the surfaces are liberated to go about their normal business. The increased entropy (disorder) of the water is favorable and drives (causes) the association of the hydrophobic surfaces.

that provides most of the driving force. As water squeezes out the hydrophobic side chains, distant parts of the protein are brought together into a compact structure. The hydrophobic core of most globular proteins is very compact, and the pieces of the hydrophobic core must fit together rather precisely.

Putting a hydrophobic group into water is difficult to do (unfavorable). Normally, water forms an extensive hydrogen-bonding network with itself. The water molecules are constantly on the move, breaking and making new hydrogen bonds with neighboring water molecules. Water has two hydrogen bond donors (the two H—O bonds) and two hydrogen bond acceptors (the two lone electron pairs on oxygen), so a given water molecule can make hydrogen bonds with neighboring water molecules in a large number of different ways and in a large number of different directions. When a hydrophobic molecule is dissolved in water, the water molecules next to the hydrophobic molecule can interact with other water molecules only in a direction away from the hydrophobic molecule. The water molecules in contact with the hydrophobic group become more organized. In this case, organization means restricting the number of ways that the water molecules can be arranged in space. The increased organization (restricted freedom) of water that occurs around a hydrophobic molecule represents an unfavorable decrease in the entropy of water.[2] In the absence of other factors, this increased organization (decreased entropy) of water causes hydrophobic molecules to be insoluble.

[2] As with most desks and notebooks, disorder is the natural state. Order requires the input of energy. Reactions in which there is an increasing disorder are more favorable. Physical chemists (and sometimes others) use the word *entropy* instead of *disorder*. There's a discussion of entropy at the end of this book.

The surface area of a hydrophobic molecule determines how unfavorable the interaction between the molecule and water will be. The bigger the surface area, the larger the number of ordered water molecules and the more unfavorable the interaction between water and the hydrophobic molecule. Bringing hydrophobic residues together minimizes the surface area directly exposed to water. Surface area depends on the square of the radius of a hydrophobic "droplet," while volume depends on the cube of the radius. By bringing two droplets together and combining their volume into a single droplet of larger radius, the surface area of the combined, larger droplet is less than that of the original two droplets. When the two droplets are joined together, some of the organized water molecules are freed to become "normal." This increased disorder (entropy) of the liberated water molecules tends to force hydrophobic molecules to associate with one another. The hydrophobic interaction provides most of the favorable interactions that hold proteins (and DNA) together. For proteins, the consequence of the hydrophobic interaction is a compact, hydrophobic core where hydrophobic side chains are in contact with each other.

VAN DER WAALS INTERACTIONS AND LONDON DISPERSION FORCES

These are very short-range interactions between atoms that occur when atoms are packed very closely to each other.

When the hydrophobic effect brings atoms very close together, van der Waals interactions and London dispersion forces, which work only over very short distances, come into play. This brings things even closer together and squeezes out the holes. The bottom line is a very compact, hydrophobic core in a protein with few holes.

HYDROGEN BONDS

Hydrogen bonding means sharing a hydrogen atom between one atom that has a hydrogen atom (donor) and another atom that has a lone pair of electrons (acceptor):

$$-C{=}O{\cdot}{\cdot}H_2O \qquad H_2O{\cdot}{\cdot}H-N- \qquad -C{=}O{\cdot}{\cdot}H-N- \qquad H_2O{\cdot}{\cdot}H_2O$$

The secondary structure observed in proteins is there to keep from losing hydrogen bonds.

A hydrogen bond is an interaction between two groups in which a weakly acidic proton is shared (not totally donated) between a group that has a proton (the donor) and a group that can accept a proton (the acceptor). Water can be both a hydrogen-bond donor and a hydrogen-bond acceptor. In an unfolded protein, the hydrogen-bond donors and acceptors make hydrogen bonds with water. Remember that the polar amino acids have groups that can form hydrogen bonds with each other and with water. The peptide bond [$-C(=O)-NH-$] that connects all the amino acids of a protein has a hydrogen-bond donor (NH) and a hydrogen-bond acceptor ($=O$). The peptide bond will form hydrogen bonds with itself (secondary structure) or with water.

Everything is just great until the hydrophobic interaction takes over. Polar peptide bonds that can form hydrogen bonds connect the amino acid side chains. Consequently, when hydrophobic residues aggregate into the interior core, they must drag the peptide bonds with them. This requires losing the hydrogen bonds that these peptide bonds have made with water. If they are not replaced by equivalent hydrogen bonds in the folded structure, this costs the protein stability. The regular structures (helix, sheet, turn) that have become known as *secondary structure* provide a way to preserve hydrogen bonding of the peptide backbone in the hydrophobic environment of the protein core by forming regular, repeating structures.

SECONDARY STRUCTURE

Secondary structure is not just hydrogen bonds.

α **Helix:**   Right-handed helix with 3.6 amino acid residues per turn. Hydrogen bonds are formed parallel to the helix axis.

β **Sheet:**   A parallel or antiparallel arrangement of the polypeptide chain. Hydrogen bonds are formed between the two (or more) polypeptide strands.

β **Turn:**   A structure in which the polypeptide backbone folds back on itself. Turns are useful for connecting helices and sheets.

Secondary structure exists to provide a way to form hydrogen bonds in the interior of a protein. These structures (helix, sheet, turn) provide ways to form regular hydrogen bonds. These hydrogen bonds are just replacing those originally made with water.

As a protein folds, many hydrogen bonds to water must be broken. If these broken hydrogen bonds are replaced by hydrogen bonds within the

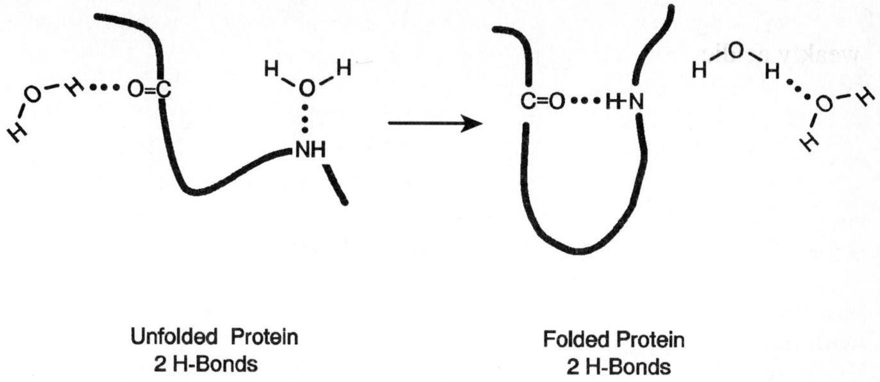

Unfolded Protein
2 H-Bonds

Folded Protein
2 H-Bonds

Figure 2-2 Solvation in Protein Folding
In an unfolded protein, water makes hydrogen bonds to all the donors and
acceptors. As the protein folds and some polar groups find themselves inside,
many of the hydrogen bonds with the solvent are replaced by hydrogen bonds
between the different donors and acceptors in the protein. Because hydrogen
bonds are being replaced rather than gained or lost as the protein folds, there is
not a large net stabilization of the protein by the hydrogen bonds.

protein, there is no net change in the number of hydrogen bonds (Fig.
2-2). Because the actual number of hydrogen bonds does not change as
the secondary structure is formed, it is often argued that hydrogen bonds
don't contribute much to the stability of a protein. However, hydrogen
bonds that form after the protein is already organized into the correct
structure may form more stable hydrogen bonds than the ones to water.
Hydrogen bonding does contribute somewhat to the overall stability of a
protein; however, the hydrophobic interaction usually dominates the
overall stability.

Small peptides generally do not form significant secondary structure
in water (there are some that do). For small peptides that do not form
stable secondary structure, there are often other favorable interactions
within the peptide that stabilize the formation of the helix or sheet structure.

The stability of secondary structure is also influenced by surrounding
structures (Fig. 2-3). Secondary structure may be stabilized by interactions between the side chains and by interactions of the side chains with
other structures in the protein. For example, it is possible to arrange the
amino acid sequence of a protein or peptide into a helix that has one face
that is hydrophobic and one that is hydrophilic. The helix wheel shown in
Fig. 2-3 illustrates how this is possible. View the helix as a long cylinder.
The peptide backbone spirals up and around the cylinder. The side chains

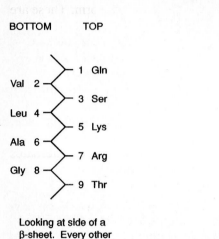

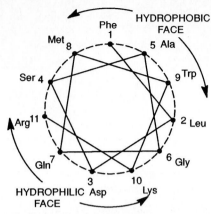

Looking at side of a β-sheet. Every other residue is on the same face of the sheet.

Looking down the axis of an α-helix. Residue sequence is numbered. The angle between residues is 360°/3.6 residues or 100°.

Figure 2-3
SECONDARY-STRUCTURE STABILIZATION is not provided by just the hydrogen bonds. On the left, you're looking at a representation of a β sheet in which the amino acid side chains alternately stick up and down. If every other side chain is hydrophobic, one side of the sheet will be hydrophobic and the other side will be hydrophilic. Interaction of the hydrophobic side with a hydrophobic region on the protein will add stability to the β sheet. On the right an α helix is shown with a hydrophobic and a hydrophilic face. Again, putting the hydrophobic face (or surface) up against another hydrophobic region of the protein will stabilize the helix. In the helix representation, there is a 100° angle (360°/3.6 residues) between residues. Side chains would stick out from the side of the cylinder defined by the helix.

of the amino acid residues point out from the helix. Each amino acid residue moves up the helix and around the helix at an angle of 100° (360°/turn = 3.6 residues/turn = 100°/residue). What you see in Fig. 2-3 is a view looking down the helix axis. The side chains are on the side of the circle (cylinder). Notice in the helix that's shown that one surface of the helix has only hydrophobic side chains, while the other side has hydrophilic side chains. This is termed an *amphipathic* helix (or *amphiphilic*, depending on whether you're a lover or a hater). With these kinds of helices, the hydrophobic face is buried in the interior while the hydrophilic face is exposed to water on the surface. There are two ways to look at this. The formation of the helix allows it to interact in a very specific way with the rest of the protein. Alternatively, you could suppose that the interaction with the rest of the protein allows the helix to form. These are equivalent ways to view things, and energetically it doesn't make any difference (see linked thermodynamic functions in Chap. 22 if you dare)— the result is that the presence of a hydrophobic and a hydrophilic side of a helix and a complementary hydrophobic region in the interior of the protein makes it more favorable to form a helix. Secondary structure can be stabilized by interactions with other parts of the protein.

β Sheets can also have a hydrophobic face and a hydrophilic face. The backbone of the β sheet is arranged so that every other side chain points to the same side of the sheet. If the primary sequence alternates hydrophobic–hydrophilic, one surface of the sheet will be hydrophobic and the other will be hydrophilic.

PROTEIN STABILITY

Protein stability is proportional to the free-energy difference between an unfolded protein and the native structure (Fig. 2-4).

It's a miracle that we're here at all. Most proteins are not very stable even though there are a large number of very favorable interactions that can be seen in the three-dimensional structure. The reason is that the favorable interactions are almost completely balanced by unfavorable interactions that occur when the protein folds. A reasonably small net protein stability results from a small net difference between two large numbers. There are lots of favorable interactions but also lots of unfavorable interactions.

Protein stability is just the difference in free energy between the correctly folded structure of a protein and the unfolded, denatured form. In the denatured form, the protein is unfolded, side chains and the peptide backbone are exposed to water, and the protein is conformationally mobile (moving around between a lot of different, random structures). The more stable the protein, the larger the free energy difference between the unfolded form and the native structure.

You can think about the energy difference in terms of an equilibrium constant if you want. For the folding reaction, the equilibrium constant K_{eq} = [native]/[denatured] is large if the protein is stable. Proteins can be denatured (unfolded) by increasing the temperature, lowering the pH, or adding detergents, urea, or guanidine hydrochloride. Urea and guanidine hydrochloride denature proteins by increasing the solubility of the hydrophobic side chains in water. Presumably these compounds,

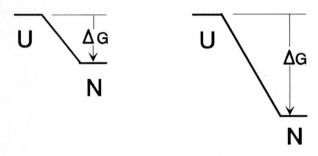

$$U \overset{\longleftarrow}{\longrightarrow} N$$

Unfolded   Folded

$$K_{eq} = \frac{[N]}{[U]}$$

More stable protein
More favorable equilibrium constant
More negative ΔG

Figure 2-4
The **FREE-ENERGY CHANGE** during a reaction like the folding of a protein is
related to how big the equilibrium constant is. For reactions that are downhill and
favorable, the free energy of the product is lower than that of the reactant. The
change in free energy (products − reactants) is less than zero (negative). Very
downhill reactions have very large equilibrium constants.

which are polar, alter water structure in some way to make it easier to
dissolve hydrophobic molecules.[3]

Protein structure (and also the interactions between proteins and
small molecules) is a compromise. It may be necessary to sacrifice a
hydrogen bond or two in order to gain two or three hydrophobic interactions. On the other hand, it may be necessary to place a hydrophobic
residue in contact with water in order to pick up a few more hydrogen

---

[3] You may have figured out from this sentence that it's not exactly known how urea and
guanidine denature proteins.

bonds in secondary structure. So it's all a compromise—a constant game of give and take. The game involves getting as many favorable interactions as you can while doing as few of the unfavorable things as possible.

FAVORABLE (GOOD) INTERACTIONS

Try to get as many of these as possible:

1. Hydrophobic interactions
2. van der Waals interactions
3. London dispersion forces
4. Hydrogen bonds
5. Charge–charge interactions

These are the favorable interactions that were discussed above. They work together to provide stabilizing interactions that hold the structure together.

UNFAVORABLE (BAD) INTERACTIONS

Avoid as many of these as possible:

1. Organizing anything into a structure (decreasing entropy)
2. Removing a polar group from water without forming a new hydrogen bond to it
3. Removing a charged group from water without putting an opposite charge nearby or putting two like charges close together
4. Leaving a hydrophobic residue in contact with water
5. Putting two atoms in the same place (steric exclusion)

There are numerous bad things (energetically speaking) that can happen when proteins fold into a three-dimensional structure. The worst thing that has to happen is that lots of covalent bonds in the protein must assume relatively fixed angles. They're no longer free to rotate as they were in the unfolded form. Protein folding requires a large loss in the conformational entropy (disorder) of the molecule. Restriction of the conformational freedom is probably the biggest unfavorable factor opposing the folding of proteins.

When a protein folds, most of the hydrophobic side chains pack into the interior. As they move into the interior, they must drag the polar amides of the polypeptide backbone with them. These backbone amides must lose contact with water and break hydrogen bonds to solvent.[4] If these hydrogen bonds that were formed with the solvent aren't replaced by new hydrogen bonds between the different polar groups that now find themselves in the interior, there will be a net loss in the number of hydrogen bonds upon folding—this is not good. Secondary structure provides a way to allow much of the polypeptide backbone to participate in hydrogen bonds that replace the ones made with water. But then there's the odd residue that just may not be able to find a suitable hydrogen-bonding partner in the folded protein. This costs energy and costs the protein stability. The same thing happens with charged residues (although they're almost always ion-paired). By the same token, it may occasionally be necessary to leave a hydrophobic group exposed to water. It may not be possible to bury all the hydrophobic residues in the interior. If not, this is also unfavorable and destabilizes the protein. All these unfavorable interactions sum up to make the protein less stable.

Don't get the impression that proteins need to be as stable as possible and that the unfavorable interactions are necessarily bad. Proteins shouldn't live forever. A good bit of metabolism is regulated by increasing and decreasing the amount of a specific enzyme or protein that is available to catalyze a specific reaction. If a protein were too stable, it might not be possible to get rid of it when necessary.

The net result of all the favorable and unfavorable interactions is that they're almost balanced. For a 100-residue protein, it is possible to estimate roughly that the sum of all the favorable interactions that stabilize the three-dimensional, native structure is on the order of -500 kcal/mol. This comes from all the favorable hydrophobic, van der Waals, hydrogen-bonding, and electrostatic interactions in the native protein. On the other hand, the sum of all the unfavorable interactions that destabilize the structure is probably near $+490$ kcal/mol. These come from conformational entropy losses (organization of the protein into a structure) and other unfavorable effects such as leaving a hydrophobic group exposed to water or not forming a hydrogen bond in the interior after having lost one that was made to water in the unfolded state. The net result is that the three-dimensional structure of a typical protein is only about -10 to -15 kcal/mol more stable than the denatured, structureless state.

---

[4] The same argument applies to polar groups on the side chains of the amino acids.

TEMPERATURE-SENSITIVE MUTATIONS

These are mutations that decrease the stability of a protein so that the denaturation temperature is near 40°C.

A single methylene group ($-CH_2-$) involved in a hydrophobic interaction may contribute as much as -1.5 to -2 kcal/mol to the stability of a protein that is only stable by -10 kcal/mol. A single hydrogen bond might contribute as much as -1.5 to -3.5 kcal/mol. If a mutation disrupts interactions that stabilize the protein, the protein may be made just unstable enough to denature near body (or culture) temperature. It might strike you as strange that we were talking above about how hydrogen bonds didn't contribute much to the net stability of proteins and now I'm telling you they contribute -1.5 to -3.5 kcal/mol. Both statements are more or less right. In the first case we were considering the folding process in which a hydrogen bond to solvent is replaced by a hydrogen bond in the folded protein—the result is a small contribution of a hydrogen bond to stability. What we're talking about now is messing up a protein by changing one amino acid for another by mutation. Here we're destroying an interaction that's present in the intact, folded protein. For any hydrogen-bonded group in the folded protein, there must be a complementary group. A donor must have an acceptor, and vice versa. Making a mutation that removes the donor of a hydrogen bond leaves the acceptor high and dry, missing a hydrogen bond. In the unfolded protein, the deserted acceptor can be accommodated by water; however, in the folded protein the loss of the donor by mutation hurts. It costs a hydrogen bond when the protein folds. The result: a loss in stability for the protein. Loss in stability means that the protein will denature at a lower temperature than before.

Temperature-sensitive mutations usually arise from a single mutation's effect on the stability of the protein. Temperature-sensitive mutations make the protein just unstable enough to unfold when the normal temperature is raised a few degrees. At normal temperatures (usually 37°C), the protein folds and is stable and active. However, at a slightly higher temperature (usually 40 to 50°C) the protein denatures (melts) and becomes inactive. The reason proteins unfold over such a narrow temperature range is that the folding process is very cooperative—each interaction depends on other interactions that depend on other interactions.

For a number of temperature-sensitive mutations it is possible to find

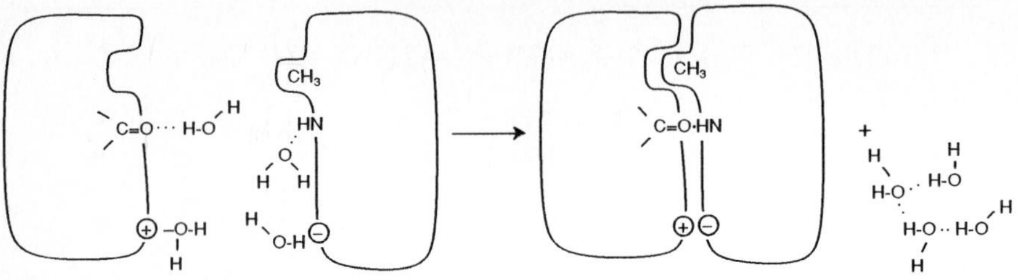

Figure 2-5
The **ASSOCIATION** of two molecules uses the same interactions that stabilize a protein's structure: hydrophobic interactions, van der Waals interactions, hydrogen bonds, and ionic interactions. To get the most out of the interaction, the two molecules must be complementary.

(or make) a second mutation in the protein that will suppress the effects of the first mutation. For example, if the first mutation decreased the protein stability by removing a hydrogen-bond donor, a second mutation that changes the acceptor may result in a protein with two mutations that is just as stable as the native protein. The second mutation is called a *suppressor mutation*.

LIGAND-BINDING SPECIFICITY

This is also a compromise (Fig. 2-5).

The specificity of the interaction of a protein with a small molecule or another protein is also a compromise. We've just said that charge–charge and hydrogen-bond interactions don't contribute a lot to the stability of a protein because their interaction in the folded protein simply replaces their individual interaction with water. The same may be said of the interaction between an enzyme and its substrate or one protein and another. However, there is a huge amount of specificity to be gained in these kinds of interactions. For tight binding, the protein and its ligand must be complementary in every way—size, shape, charge, and hydrogen-bond donor and acceptor sites.

Both the protein and the ligand are solvated by water when they are separated. As the two surfaces interact, water is excluded, hydrogen bonds are broken and formed, hydrophobic interactions occur, and the protein and ligand stick to each other. As in protein folding and for the same reasons, the hydrophobic interaction provides much of the free energy for the association reaction, but polar groups that are removed

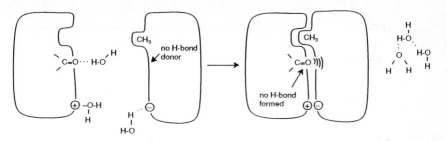

Figure 2-6
SPECIFICITY in the association of two proteins or a protein and a small molecule results from the requirement that the two interacting molecules must be complementary—complementary in charge, hydrogen bonding, and hydrophobic patches as well as shape. If any of the possible interactions are not satisfied, the strength of the interaction suffers.

from water by the interaction must find suitable partners in the associated state.

Consider what happens when a nonoptimal ligand binds to the protein. The binding of this modified ligand is much weaker not because it's not the right size to fit into the protein binding site, but because the complementary group on the protein loses a favorable interaction with water that is not replaced by an equally favorable interaction with the ligand (Fig. 2-6).

As with the formation of secondary structure, the multiple, cooperative hydrogen bonds that can be formed between the ligand and the protein may be stronger and more favorable than hydrogen bonds that the ligand might make to water. Hydrogen bonding may, in fact, make some contribution to the favorable free energy of binding of ligands to proteins.

GLOBAL CONCLUSION

Now that you understand the basis for the interactions between functional groups in water, you also understand the basis for most interactions: DNA–DNA, DNA–RNA, DNA–protein, RNA–protein, protein–protein, protein–ligand, enzyme–substrate (Get the picture?), antibody–antigen, protein–chromatography column—it's all the same stuff.

DNA-RNA STRUCTURE

·

DNA Structure

DNA Stability

RNA Secondary Structure

·   ·   ·   ·   ·   ·   ·   ·   ·   ·   ·   ·

DNA STRUCTURE

Double helix
A = Adenine   = purine
T = Thymine   = pyrimidine (DNA only)
G = Guanine   = purine
C = Cytosine = pyrimidine
U = Uracil   = pyrimidine (RNA only)
AT/GC base pairs
Antiparallel strands
Major groove–minor groove
A-, B-, and Z-DNA

The two complementary strands of the DNA double helix run in antiparallel directions (Fig. 3-1). The phosphodiester connection between individual deoxynucleotides is directional. It connects the 5′-hydroxyl group of one nucleotide with the 3′-hydroxyl group of the next nucleotide. Think of it as an arrow. If the top strand sequence is written with the 5′ end on the left (this is the conventional way), the bottom strand will have a complementary sequence, and the phosphate backbone will run in the opposite direction; the 3′ end will be on the left. The antiparallel direc-

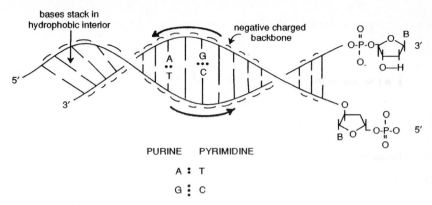

PURINE   PYRIMIDINE

A : T

G : C

Figure 3-1   Structural Features of DNA

tionality of DNA is an important concept (i.e., it always appears on exams). Either of the two strands could be written on top (just rotate the paper by 180°), but if the DNA codes for a protein, the top strand is usually arranged so that it matches the sequence of the RNA that would be made from the DNA (see below). In Fig. 3-2, you're looking at a base pair as it would be seen from above, looking down the helix axis. The DNA double helix has two grooves—the major and the minor. If the helix were flat, the major and minor grooves would correspond to the two different flat surfaces represented by the front and back of the flat sheet. The major and minor grooves are of different size because the two strands come together so that the angle between corresponding points on the phosphate backbone is not 180°. Many of the sequence-specific interactions of proteins with DNA occur along the major groove because the bases (which contain the sequence information) are more exposed along this groove.

The structures shown above are for B-form DNA, the usual form of the molecule in solution. Different double-helical DNA structures can be formed by rotating various bonds that connect the structure. These are termed different conformations. The A and B conformations are both right-handed helices that differ in pitch (how much the helix rises per turn) and other molecular properties. Z-DNA is a left-handed helical form of DNA in which the phosphate backbones of the two antiparallel DNA strands are still arranged in a helix but with a more irregular appearance. The conformation of DNA (A, B, or Z) depends on the temperature and salt concentration as well as the base composition of the DNA. Z-DNA appears to be favored in certain regions of DNA in which the sequence is rich in G and C base pairs.

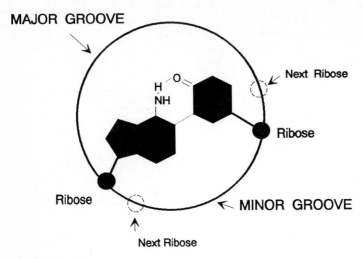

Figure 3-2
DNA has a **MAJOR AND MINOR GROOVE** because the bases attach at an angle
that is not 180° apart around the axis of the helix. The major groove has more of
the bases exposed. Sequence-specific interactions with DNA often occur along
the major groove. Since the helix is right-handed, the next ribose shown is above
the last one.

DNA STABILITY

Melting is denaturation.
Annealing is renaturation.
Hydrophobic stacking provides stability.
Intercalating agents stack between bases.

STABILITY INCREASED BY
Decreased temperature
Increased GC content (three hydrogen bonds)
Increased salt (ionic strength)

The DNA double helix is stabilized by hydrophobic interactions
resulting from the individual base pairs' stacking on top of each other in
the nonpolar interior of the double helix (Figs. 3-1 and 3-2). The hydrogen
bonds, like the hydrogen bonds of proteins, contribute somewhat to the
overall stability of the double helix but contribute greatly to the specificity

for forming the correct base pairs. An incorrect base pair would not be able to form as many hydrogen bonds as a correct base pair and would be much less stable. The hydrogen bonds of the double helix ensure that the bases are paired correctly.

The double helix can be denatured by heating (melting). Denatured DNA, like denatured protein, loses its structure, and the two strands separate. Melting of DNA is accompanied by an increase in the absorbance of UV light with a wavelength of 260 nm. This is termed *hyperchromicity* and can be used to observe DNA denaturation. DNA denaturation is reversible. When cooled under appropriate conditions, the two strands find each other, pair correctly, and reform the double helix. This is termed *annealing*.

The stability of the double helix is affected by the GC content. A GC base pair has three hydrogen bonds, while an AT base pair has only two. For this reason, sequences of DNA that are GC-rich form more stable structures than AT-rich regions.

The phosphates of the backbone, having a negative charge, tend to repel each other. This repulsion destabilizes the DNA double helix. High ionic strength (high salt concentration) shields the negatively charged phosphates from each other. This decreases the repulsion and stabilizes the double helix.

Intercalating agents are hydrophobic, planar structures that can fit between the DNA base pairs in the center of the DNA double helix. These compounds (ethidium bromide and actinomycin D are often-used examples) take up space in the helix and cause the helix to unwind a little bit by increasing the pitch. The pitch is a measure of the distance between successive base pairs.

RNA SECONDARY STRUCTURE

Stem: A stretch of double-stranded RNA
Loop: A loop of RNA
Hairpin loop: A very short loop
Pseudoknot: Interaction between one secondary structure element and another part of the same RNA molecule

---

RNA is often depicted as a single-stranded molecule. However, in many RNAs, internal complementarity may result in secondary (and tertiary) structure in which one part of the RNA molecule forms a double-

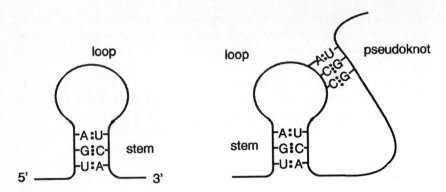

Figure 3-3   RNA Secondary Structure
A single molecule of RNA often contains segments of sequence that are complementary to each other. These complementary sequences can base-pair and form
helical regions of secondary structure. Interactions between the secondary structures give RNA a significant folded, three-dimensional structure.

stranded region with another part of the same molecule. There are usually
a number of mismatches in these structures. Names have been given to
some of these structural features (Fig. 3-3).

EXPRESSION OF GENETIC INFORMATION

·

Information Metabolism

Directions and Conventions

DNA Replication

Recombination

Regulation of Information Metabolism

Transcription

Regulation of Transcription

Translation

Use of High-Energy Phosphate Bonds during Translation

· · · · · · · · · · · · ·

INFORMATION METABOLISM

DNA → RNA → protein → structure.

Information metabolism provides a way to store and retrieve the information that guides the development of cellular structure, communication, and regulation. Like other metabolic pathways, this process is highly regulated. Information is stored by the process of DNA replication and meiosis, in which we form our germ-line cells. These processes are limited to specific portions of the cell cycle. Information is retrieved by the transcription of DNA into RNA and the ultimate translation of the signals in the mRNA into protein.

Regulation of information metabolism occurs at each stage. The net result is that specific proteins can be made when their activities are needed.

DIRECTIONS AND CONVENTIONS

The 5′ end of the top (sense) strand is on the left.
Top strand = RNA sequence.
Decoded RNA sequence in 5′ to 3′ direction gives protein sequence in N to C direction (Fig. 4-1).

DNA is a double-stranded molecule in which the structure of the second strand can be deduced from the structure of the first strand. The second strand is complementary (A's in the first strand match T's in the

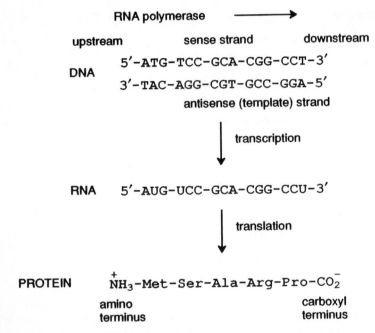

Figure 4-1
DIRECTIONALITIES in the flow of information from DNA to RNA to protein. All new DNA or RNA chains grow by adding new nucleotides to a free 3′ end so that the chain lengthens in the 5′ to 3′ direction. Protein is made by reading the RNA template starting at the 5′ end and making the protein from the N to the C terminus.

second and G's match C's) and runs in the opposite direction. This means that you don't have to write both strands to specify the structure—one will do.

When you see a sequence written with only one strand shown, the 5' end is written on the left. Usually this sequence is also identical to that of the RNA that would be made from this piece of DNA when transcribed left to right. The DNA strand that has the same sequence (except U for T) as the RNA that is made from it is called the *sense strand*. The sense strand has the same sequence as the mRNA. The antisense strand serves as the template for RNA polymerase.

The protein synthesis machinery reads the RNA template starting from the 5' end (the end made first) and makes proteins beginning with the amino terminus. These directionalities are set up so that in prokaryotes, protein synthesis can begin even before the RNA synthesis is complete. Simultaneous transcription-translation can't happen in eukaryotic cells because the nuclear membrane separates the ribosome from the nucleus.

When writing protein sequences, you write the amino terminus on the left. If you have to use the genetic code tables to figure out a protein sequence from a DNA sequence, it is not necessary to write down the complementary RNA sequence first; it's the same as that of the sense strand (the one on top).

DNA REPLICATION

Origin is the beginning.
New chains grow **5' to 3'**.
Bidirectional synthesis.
Leading strand = continuous synthesis.
Lagging strand = discontinuous synthesis.
Order of action:
  Unwinding proteins.
  Single-strand binding proteins.
  Primase makes RNA primer.
  DNA polymerase makes DNA.
  RNAse H removes RNA primer.
  DNA polymerase fills in gaps.
  DNA ligase joins gaps.

Keeping your direction in mind is never a bad idea, but with replication, transcription, and translation it's absolutely essential—these types

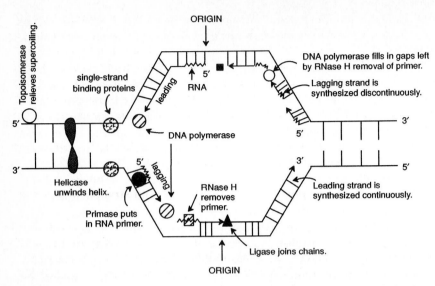

Figure 4-2
DNA REPLICATION begins at a defined origin, is bidirectional, and is semicon-servative (one new chain, one old chain in daughter DNA), and chain growth occurs in the 5′ to 3′ direction.

of questions are just too easy to write, and you'll see them for certain (Fig. 4-2).

All DNA polymerases are single-minded—they can do it only one way. Each dNTP (deoxynucleoside triphosphate) is added to the 3′-OH group of the growing chain so that all chains grow from the 5′ end in the direction 5′ to 3′. Since strands are antiparallel, the template strand is read in the 3′ to 5′ direction. This is true of both DNA and RNA synthesis. Most of what you need to know about DNA replication can be summarized in a single picture.

To remember the order in which things happen, you must understand the structure of chromosomal DNA, and directions. Then it's just a matter of developing a mechanical picture of how things must be done in order to get access to the information and make a copy. Chromosomal DNA is normally packaged around histones. At unique DNA sites called *origins of replication,* unwinding proteins (helicases) unwind the helix in an ATP-dependent manner. Single-strand binding proteins then bind to and stabilize the single-stranded DNA regions to keep them single-stranded.

In addition to a template (a DNA sequence that specifies the order in which the nucleotides will be jointed), DNA polymerase requires a primer. A primer is a short piece of DNA or RNA that is complementary to the template and has a free 3' end onto which the growing strand can be elongated. DNA polymerase can't prime itself—it must have a 3' end to get started. In eukaryotic DNA replication, these primers are actually RNA. A special RNA polymerase (primase) puts them in, and later the RNA primer is removed by a specific ribonuclease (RNAse H), and the gaps are filled in by DNA polymerase. This may be a mechanism to enhance the fidelity of DNA replication.

DNA replication proceeds in both directions from the replication origin (bidirectional), which means you need to form two sets of replication complexes. Each replication complex moves away from the origin (in opposite directions), unwinding and replicating both strands at each replication fork.

The strands of DNA are wound around each other like the strands of a rope. As the strands are pulled apart during the movement of a replication fork, this unwinding tends to make the ends of the DNA turn (imagine unwinding the strands of a rope). Since the DNA is very long, twisted, and wrapped around histones, the DNA really can't turn—its ends are rather tied down. The unwinding of one region around the replication forks introduces strain into the regions of DNA that are still double-stranded, tending to make them wind tighter. This is called *supercoiling* of the DNA. Proteins (topoisomerases) are present to relieve the strain associated with helix unwinding by nicking and rejoining the DNA in the double-stranded regions.

At each replication fork there are two DNA polymerase complexes. As the double-stranded DNA is unwound, two template strands are exposed. One of the templates can be replicated in a continuous fashion by DNA polymerase since a continuous synthesis of new strands can occur in the 5' to 3' direction as the template strand is exposed. Since all growing chains must be synthesized in the 5' to 3' direction, the lagging chain must be continuously reinitiated as new template is exposed. The lagging strand is then synthesized discontinuously, in pieces that must be joined together later.

After synthesis, the RNA primers must be removed, gaps filled in, and the strands joined to give a linear, duplex DNA. New histones are added to the lagging strand (which is now a duplex) while the old histones remain with the leading strand. As the smoke slowly clears, we have a copy of the original DNA.

Since all DNA polymerases require a primer and work only in the 5' to 3' direction, there's a problem with replicating the 5' ends of the DNA. If an RNA primer has to be laid down and later removed, these ends can't get replicated. For bacteria with a circular genome, this isn't a problem. Eukaryotes have specialized structures called *telomeres* at the ends of the chromosome to solve this problem. The exact details aren't known, but telomeres at the ends of each chromosome consist of a larger number (3000 to 12,000 base pairs) of a tandem (side by side) repeat of a G-rich sequence, $(TTAGGG)_n$, in human DNA. A specialized enzyme, telomerase, that also contains an RNA cofactor is responsible for replication of at least one strand of these telomeric sequences.

RECOMBINATION

Recombination rearranges genetic information by breaking and joining DNA.

Homologous: Two DNA sequences that are very similar or identical. Homologous recombination occurs between two genes that have very similar or identical sequences.

Nonhomologous: Two DNA sequences that are very different. Nonhomologous recombination can occur between two unrelated genes.

Aligned: Recombination occurs between the same genes and at the same location within each gene. Gene order is not altered.

Nonaligned: Recombination occurs between two different genes. The order of genes is altered by nonaligned recombination.

There are lots of ways of moving genetic information around. All contribute to genetic diversity in the population. The result of recombination can be pictured as breaking two DNA strands into two pieces, swapping the ends, and rejoining. At the level of the individual strands, it's a little more complicated, but for our purposes it's good enough.

Recombination can occur in regions of sequence homology. If these homologous regions correspond to the same position in the same gene, this is an aligned recombination (also called recombination with equal crossing over). If all the genes on the two chromosomes are the same, then recombination won't have any effect. But if one of the genes contains a mutation, recombination results in two new chromosomal structures in which different genes are linked to the site of the mutation. Note that in recombination between two chromosomes, no information is actually lost—all the DNA ends up somewhere. However, each offspring receives only one of the two new chromosomes (Fig. 4-3).

If recombination occurs between two regions of homology that are in

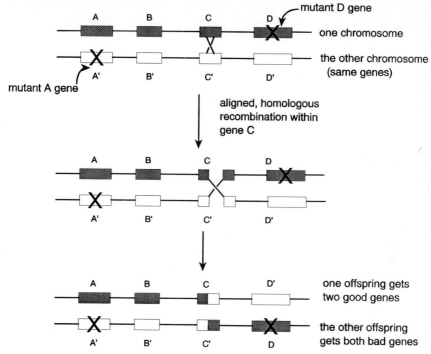

Figure 4-3
ALIGNED, HOMOLOGOUS RECOMBINATION swaps information between the
same genes on two copies of the same chromosome. Genes are not lost or
duplicated, nor is their order changed. Different combination of specific alleles
(copies of same gene) does occur.

different genes (unaligned recombination or unequal crossing over), individual genes can be duplicated or lost in the resulting daughter DNA. A
good example is the globin gene family. There are several α- and β-globin
genes that share some sequence homology. If recombination occurs between two similar (but not identical) genes, the resulting DNA will have
been rearranged so that one progeny is a gene or two short while the other
offspring has a few too many. Again, no DNA has actually been lost; it's
just been redistributed between offspring (Fig. 4-4).

Gene deletion may cause genetic disease if the gene product is
essential, and gene duplication, which creates an extra copy of the gene,
can be used to help create new genes by mutation. If you've got two
copies of a gene, you can afford to fool around changing one of them, and
maybe you'll invent a new and improved gene in the process.

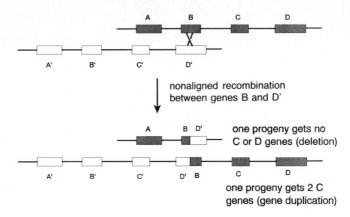

Figure 4-4   Nonaligned Homologous Recombination
Genes may be duplicated or deleted when recombination occurs between two different genes on the two copies of the same chromosome. Recombination can occur between two regions of two different genes that have some sequence homology.

During the generation of genes that direct the synthesis of antibody molecules, recombination within the same chromosome is used to bring distant segments of the gene together and to generate the diversity of recognition sites that allow different antibodies to recognize different antigens. Immunoglobulins consist of two copies of a light chain and two copies of a heavy chain. The heavy and light chains combine to generate the antigen-recognition site. The genes for the different parts of the light chain are arranged in three different clusters: a large number of gene segments for the variable regions of the light chains, a series of joining genes (J), and the constant region. A given variable region is joined to the constant region by a nonaligned recombination that deletes the DNA between the two points of recombination. A similar mechanism is used in making the heavy-chain gene [except there's another type of segment (D) and a few more types of constant regions]. These genetic rearrangements within the same piece of DNA actually cause DNA to be lost. Once the recombination is done it's done, and this cell and its offspring are committed to producing one specific light-chain protein. If the antibody made by a specific cell actually recognizes something foreign, the cell is saved and copied; if not, the cell dies. The large number of different antibody-recognition sites is made possible by the random joining of one of the many variable (V) segments to one of the joining segments by recombination (Figs. 4-5 and 4-6).

REGULATION OF INFORMATION METABOLISM

Inducible: Genes turned on by the presence of a substrate for a catabolic (degradative) pathway.

Repressible: Genes turned off by the presence of a product of a biosynthetic pathway.

Positive regulators (enhancers):   Turn on transcription when a specific effector protein binds to a specific enhancer sequence in the DNA.

Negative regulators (repressors):   Turn off transcription when a specific effector protein binds to a specific repressor sequence in the DNA.

One way to control how much of something a cell uses or makes is to control the levels of the enzymes that are required to metabolize it (Fig. 4-7). Whether or not transcription happens is controlled by the binding of specific proteins to the DNA. When they bind to DNA, these proteins can either help or hinder the transcription process. *Positive* and *negative* refer only to the effect a protein has when it binds to the DNA. A positive effect is when the protein binds to the DNA and turns on the transcription of the gene. A negative effect is when the binding of the protein to the DNA turns off transcription.

Inducible or *repressible* refers to the type of response the system makes to the presence of a metabolite. Inducible genes are turned on when they sense the presence of a metabolite. Usually, this means that the metabolite is a precursor of something the cell needs. If the precursor is present, inducible genes are turned on to metabolize it. Repressible genes are turned off by the presence of a metabolite. These genes are usually involved in the synthesis of the metabolite. If the cell has enough of the metabolite, the pathway is turned off (repressed). If the metabolite is not present, the pathway is turned on.

Operons are clusters of genes located next to each other. The proteins they make are usually required at the same time and for the same overall function. The transcription of genes in an operon is regulated by a common regulatory site(s) on the DNA. Inducible or repressible operons may be created by either positive or negative regulatory elements. The concepts of inducible/repressible and positive/negative control are related but independent. There are then two possibilities for regulation of inducible pathways. If a regulatory protein binds to DNA when it senses

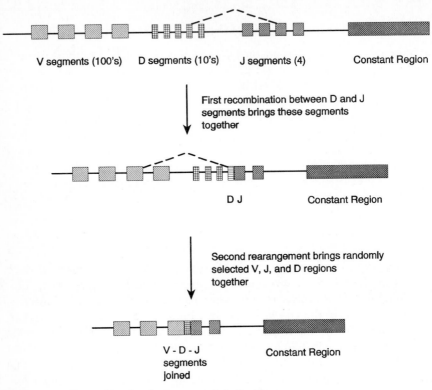

Figure 4-5   Recombination in Immunoglobulin Genes
Recombination is used to randomly combine a variable and two joining segments
of the immunoglobulin heavy-chain genes. These rearrangements generate a new
DNA that codes for an immunoglobulin heavy chain with a single antigenic
specificity. A later recombination joins the selected VDJ region to an appropriate
constant-gene segment. Similar rearrangements are used to generate the light
chain.

the metabolite and then activates transcription, this is a positive way of
inducing RNA synthesis. An inducible gene can also function by negative
regulation. If a regulatory protein binds to DNA and shuts off transcription when the metabolite is absent, and the protein is released from the
DNA when it binds the metabolite, the net effect is the same (increased
transcription). Inducible genes can be regulated by either positive or
negative effectors. There are also two ways to have repressible genes
using positive and negative regulation.

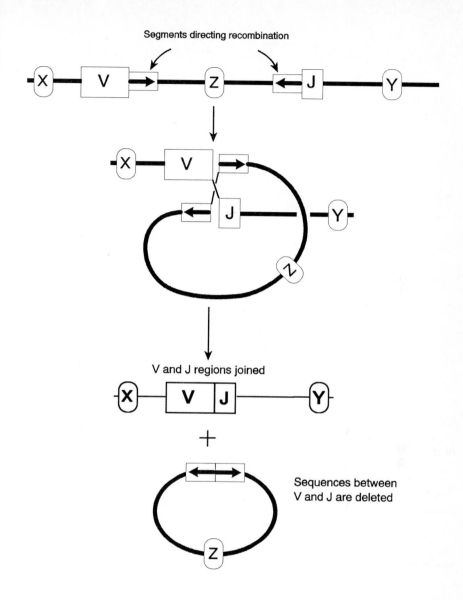

Figure 4-6
The **RECOMBINATION THAT JOINS** the V, D, and J gene segments of the
immunoglobulin heavy chain occurs between specific regions that precede and
follow the V, D, and J regions. Intragene recombination between these regions
results in deletion of the intervening DNA and joining of the two segments.

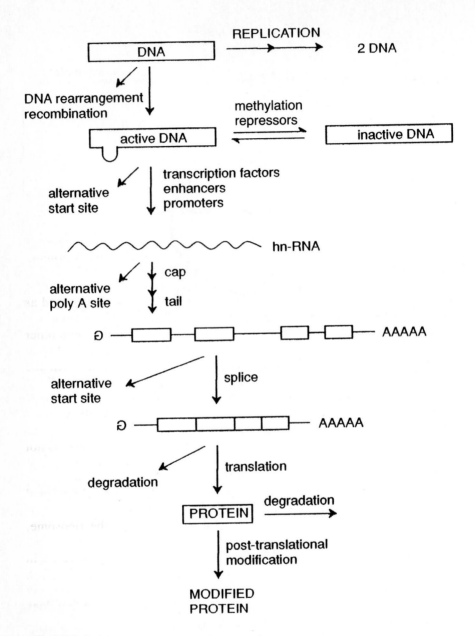

Figure 4-7
REGULATION OF INFORMATION FLOW from DNA to RNA to protein. Every aspect of the process is controlled, and alternatives are available that affect which information is expressed at what time.

TRANSCRIPTION

RNA polymerase uses the antisense strand of DNA as a template.
RNA is synthesized in the 5′ to 3′ direction.
The 5′ end is capped with inverted 7-methyl-G.
Poly(A) tail is added.
Introns are spliced out and exons joined.
RNA is exported from the nucleus.
RNA is translated into protein.
RNA is degraded.

RNA polymerase makes a copy of the sense strand of the DNA using the antisense strand as a template (Fig. 4-8). The sequence of the primary transcript is the same as that of the sense strand of the DNA. RNA polymerase needs no primer—only a template. Either of the two DNA strands can serve as the template strand. Which DNA strand is used as the template depends on the direction in which the gene is transcribed. Within the genome, some genes are transcribed left to right while other genes in the same chromosome are transcribed right to left. The direction depends on which strand actually contains the signals that form the binding site for RNA polymerase (the *promoter*). Regardless of the direction of transcription, the new RNA strand is synthesized in the 5′ to 3′ direction and the antisense strand is read in the 3′ to 5′ direction.

After synthesis, the primary transcript (hnRNA—for *heterogeneous nuclear)* is capped on the 5′ end with an inverted G residue. The G is not actually backward or inverted; the *inverted* refers to the fact that the 5′ end is capped by forming a phosphate ester between the 5′ end of the DNA and the 5′-triphosphate of 7-methyl-GTP rather than the normal 5′—3′ bond. This stabilizes the message against degradation from exonucleases and provides a feature that is recognized by the ribosome. Next the message is tailed on the 3′ end with a stretch of A's of variable length (100 to 200 nucleotides). There is not a corresponding set of T's in the DNA template. Poly(A) addition requires a sequence (AAUAAA) in the RNA that helps direct the cleavage of the transcript and the addition of the poly(A) tail by poly(A) polymerase, an RNA polymerase that does not use a template.

To make mRNA, the primary transcript must be spliced to bring the protein-coding sequences (exons) together and to remove the intervening sequences (introns). The splice signals consist of a 5′ and a 3′ set of sequences that are always found at splice junctions. However, this is

Alternative Splicing

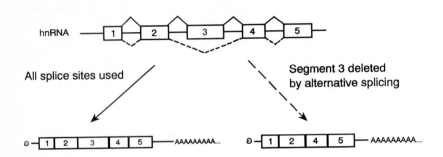

Alternative Tailing

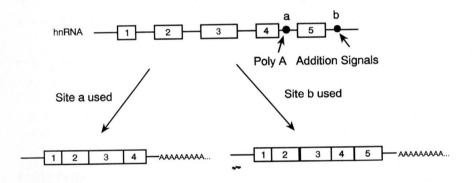

Figure 4-8
ALTERNATIVE SPLICING OR ALTERNATIVE USE OF MULTIPLE POLY(A) SITES can be used to generate an RNA (and protein) that is missing a portion of the information present in the gene. These mechanisms are useful in generating two proteins from the same gene. A soluble and a membrane-bound form of the same protein can be made from the same RNA by simply splicing out or skipping the membrane-anchor sequences during RNA processing.

generally believed to provide too little information to recognize a splice site specifically and correctly. Some sequences in the intron are also important.

After synthesis, the mRNA exits the nucleus through a nuclear pore and proceeds to the ribosome for translation into protein. Competing with export and translation is the process of message degradation by cellular ribonucleases. The competition between degradation and translation provides another mechanism to regulate the levels of individual messages.

REGULATION OF TRANSCRIPTION

Template availability
Methylation
Exposure of DNA
Attraction for RNA polymerase
Promoter: TATA, CAT
Transcription factors
Alternative Poly(A) tailing
Alternative splicing
Alternative translation start

As goes RNA, so goes protein. Higher levels of mRNA are associated with higher levels of the encoded protein. There is a definite need to regulate the amounts of different proteins during development, differentiation, and metabolism, so there are a lot of controls on the synthesis and degradation of RNA. Control of some sort is exerted at virtually every step of mRNA synthesis.

• **TEMPLATE AVAILABILITY:**   The DNA template must be available. This may be controlled by DNA methylation, histone arrangement on the DNA, and interactions of the DNA with the nuclear matrix (a catch phrase for a bunch of protein that's always found in the nucleus).

• **ATTRACTION FOR RNA POLYMERASE:**   RNA polymerase binds to DNA at specific sites (called promoters) to initiate transcription. A major site is the TATA box (named for the consensus sequence[1] that is often found there) that is located about 25 nucleotides upstream (on the 5' side) of the translation start site. Not all genes have TATA boxes, and not all promoters have the same efficiency—some are better than others. For many genes, there are other DNA sequences that regulate transcription by binding specific proteins (transcription factors). These transcription factors (enhancers and repressors) may help or hinder transcription. The transcription factor binding sites may be located at varying distances from

---

[1] Consensus sequences are sequences that agree with each other more or less. Often there are a few differences found among the different genes that might have a given consensus sequence. It can be viewed as an "average" sequence.

the transcription start site, and a given promoter region may be affected by more than one of these enhancer or repressor sites. The binding of transcription factors to a specific site on the DNA regulates the transcription by enhancing or inhibiting the formation of the complex structure that is required to initiate transcription. The rules of this regulatory game are not totally sorted out. Transcription factor binding is important in the tissue-specific expression of an mRNA, the regulation of expression during development, and who knows what else.

Eukaryotes have a specific signal for termination of transcription; however, prokaryotes seem to have lost this mechanism. Once started, RNA polymerase keeps going, making a primary transcript [pre-mRNA or hnRNA (for *heterogeneous nuclear*)] until far past the end of the final mRNA message.

• **POLY(A) TAILING:** Most RNAs coding for protein are poly(A)-tailed. Having a poly(A) tail helps direct the RNA to the cytoplasm and may increase the stability of the message. One mechanism of regulation of transcription involves the alternative use of different poly(A) addition sites. Some genes have more than one poly(A) addition signal. Which signal is used can depend on the type of cell or the stage of development, or it can be used to make two kinds of protein from the same message. Alternative poly(A) addition site usage has the same effect as alternative splicing, except that it deletes terminal exons from the message and creates proteins with different COOH-terminal sequences.

• **ALTERNATIVE SPLICING:**   Most primary transcripts must be spliced to connect the proper exons. Some genes contain alternative splice sites that can be used to bring two different exons together and make different gene products depending on need. Alternative splicing changes the sequence of the actual protein that's made. It's useful for making two proteins that share a common sequence. For example, during immunoglobulin synthesis, IgM is made in two forms. One has a membrane-spanning domain so that the IgM with its antigen-recognition site is anchored to the cell plasma membrane. The other form simply lacks the membrane anchor and is secreted in a soluble form. These two forms of the IgM molecule are generated by using alternative splice sites. If the membrane-spanning region is spliced out, the protein loses the ability to bind to the membrane.

• **ALTERNATIVE START SITES:**   If all of the above didn't provide enough diversity, some messages contain two AUG initiation codons separated by some intervening information. Protein synthesis can initiate at either site. This is useful for making proteins with or without NH_2-terminal signal sequences.

TRANSLATION

Translation reads the RNA template in the 5' to 3' direction.
The amino terminus is synthesized first.
AUG = start = Met in eukaryotes and fMet in prokaryotes.

Protein synthesis (translation) is a two-component system—a system for activating individual amino acids into a chemically reactive form and a system that directs exactly which amino acid is to be used when (Fig. 4-9).

Activation of individual amino acids occurs in the synthesis of aminoacyl tRNA. This process burns two ATP equivalents (forms pyrophosphate and AMP) and connects a specific amino acid to a specific tRNA.

$$tRNA_{Phe}-3'-CCA-OH \quad + \quad {}^+NH_3-Phe-CO_2^-$$

$$\downarrow$$

$$tRNA_{Phe}-3'-CCA-O-CO-Phe-NH_3^+ + AMP + PP_i$$
aminoacyl-tRNA

The tRNA synthetases may provide a check to make sure that the correct amino acid has been attached to the correct tRNA. If an incorrect amino acid is attached to the tRNA, it will be incorporated into the protein at the position specified by the identity of the tRNA. At least some of the aminoacyl tRNA synthetases have a "proofreading" function that hydrolyzes any incorrect aminoacyl tRNAs (for example, a Val residue attached to an Ile tRNA).

Each tRNA has a different sequence at the anticodon loop that is complementary to the codon sequence in the RNA. The recognition structure that is formed is analogous to double-stranded, antiparallel DNA. If the codon (in the RNA) is GCA (written 5' to 3'), the anticodon loop in the tRNA would have the sequence UGC (again written to 5' to 3').

There are 64 different three-letter codons, but we don't have to have 64 different tRNA molecules. Some of the anticodon loops of some of the tRNAs can recognize (bind to) more than one codon in the mRNA. The anticodon loops of the various tRNAs may also contain modified bases that can read (pair with) multiple normal bases in the RNA. This turns out to be the reason for the "wobble hypothesis," in which the first two letters of a codon are more significant than the last letter. Look in a codon table and you'll see that changing the last base in a codon often doesn't

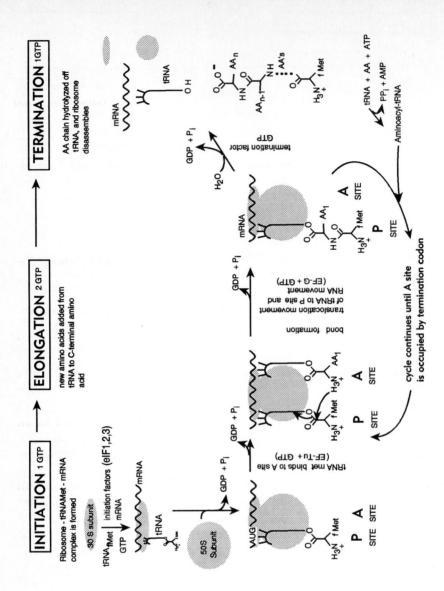

Figure 4-9 Translation

change the identity of the amino acid. A tRNA that could recognize any base in codon position 3 would translate all four codons as the same amino acid. If you've actually bothered to look over a codon table, you realize that it's not quite so simple. Some amino acids have single codons (like AUG for Met), some amino acids have only two codons, and some have four.

After attachment of amino acids to tRNA, the amino acids are assembled beginning with the amino terminus and proceeding in the direction of the carboxy terminus. The ribosome is the machinery that translates the mRNA into protein. The ribosome is a very complex protein that contains ribosomal RNA as a functional and structural component. The ribosome assembles around the mRNA, and the cap and other signals allow alignment of the mRNA into the correct position. The initial assembly of the mRNA into the ribosome requires association of the small ribosomal subunit with an initiator tRNA (Met or fMet). "Small" is a misstatement, because the small ribosomal subunit is a large, complex assembly of numerous smaller proteins—it's just smaller than the *large* subunit. This association requires a specific initiation factor and the hydrolysis of GTP. The reactions of translation are driven by the hydrolysis of GTP, not ATP. Throughout the process, elongation factors come and go, GTP gets hydrolyzed, and finally the completed protein is released from the ribosome.

Several key concepts are worth remembering. GTP is used as an energy source for translation, but ATP is used to form the aminoacyl-tRNA. The ribosome effectively has two kinds of tRNA binding sites. Only tRNAMet can bind to the P (for *peptide*) site, and this only occurs during the initial formation of the functional ribosome (initiation). All other aminoacyl-tRNAs enter at the A (for *amino acid*) binding site. After formation of the peptide bond (this doesn't require GTP hydrolysis), the tRNA with the growing peptide attached is moved (translocated) to the other site (this does require GTP hydrolysis).

USE OF HIGH-ENERGY PHOSPHATE BONDS DURING TRANSLATION

Four high-energy phosphates are used for each amino acid that is incorporated into a protein.

How many high-energy phosphate bonds are required for the synthesis of a protein from amino acids during translation?

ENERGY REQUIREMENTS FOR THE SYNTHESIS OF A 100-RESIDUE PROTEIN:

100 aminoacyl-tRNAs	(2 P each)	200
Initiation complex	(1 P each)	1
99 tRNAs binding to A site	(1 P each)	99
99 peptide bonds	(0 P each)	0
99 translocations (A to P)	(1 P each)	99
1 termination (hydrolysis)	(1 P each)	1
Total per 100 amino acids		400

Bottom line: 4 high-energy phosphates used per amino acid incorporated into a protein.

RECOMBINANT-DNA METHODOLOGY

·

Restriction Analysis

Gels and Electrophoresis

Blotting

Restriction Fragment–Linked Polymorphism

Cloning

Sequencing

Mutagenesis

Polymerase Chain Reaction

·   ·   ·   ·   ·   ·   ·   ·   ·   ·   ·   ·

Much of what we know about the regulation of information flow (gene expression) has been made possible by the ability to manipulate the structures of DNA, RNA, and proteins and see how this affects their function. The ability to manipulate DNA (recombinant-DNA methods) has generated a new language filled with strange-sounding acronyms that are easy to understand if you know what they mean but impossible to understand if you don't. Understand?

RESTRICTION ANALYSIS

Restriction enzymes are sequence-specific endonucleases that cut double-stranded DNA at specific sites.

Most useful restriction enzymes cut DNA at specific recognition sites, usually four to six nucleotides in length. There can be multiple restriction sites for a single endonuclease within a given piece of DNA, there can be only one (a unique restriction site), or there can be none. It all depends on the sequence of the specific piece of DNA in question.

Cutting with restriction endonucleases is very useful for moving specific pieces of DNA around from place to place. It's also a useful way to name pieces of DNA. For example, a piece of DNA that is cut from a bigger piece of DNA is often named by size and given a surname that corresponds to the two restriction enzymes that did the cutting—the 0.3-kb EcoRI-BamHI fragment. Restriction enzymes themselves are named for the bacterial strains from which they were initially isolated.

A restriction map shows the location of restriction sites in a given DNA sequence.

When digested with two (or more) restriction enzymes at the same time, most large pieces of DNA give a specific pattern of different-sized DNA fragments depending on the distance separating the different cleavage sites. These different fragments can then be separated by size on an agarose gel. By working backward (biochemists are good at this) from the sizes of the different DNA fragments, it is possible to construct a map that locates the different restriction sites along a given piece of DNA. For example, if we cut the 3.6-kb piece of DNA in Fig. 5-1 with SmaI, we would see two bands on the agarose gel—1.9 and 1.7 kb. This would tell us that the SmaI site is very near the middle of the fragment. We could start constructing our map by putting the 1.7-kb fragment on the left side or the right side—it doesn't matter, and we can't know which is right (or left). In Fig. 5-1, the DNA is arbitrarily put down with the smaller fragment on the right. If we cut with BamHI, we get fragments that are 0.9 and 2.7 kb. Again we wouldn't know whether to put the BamHI site on the right or left of the map, but here it does matter because we already have the SmaI site on the map. The way to decide where to put the BamHI site is to cut with both BamHI and SmaI. Let's say that you get fragments of 0.9, 1.0, and 1.7 kb. Notice that the 1.7-kb fragment is the same size as in the digest with SmaI alone. This tells you that the BamHI site is in the 1.9-kb SmaI fragment, that is, on the left side of our map. By going through this kind of reasoning over and over, it is possible to construct a map of restriction sites along your piece of DNA.

Restriction enzymes that recognize a specific sequence of five nucleotides should cut the DNA, on average, every 4^5 base pairs (this is the frequency with which a given sequence of five nucleotides would occur by chance), or every 1024 base pairs. As a result, the average size of most

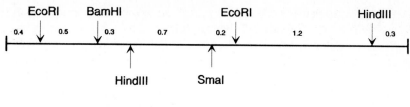

Restriction Map

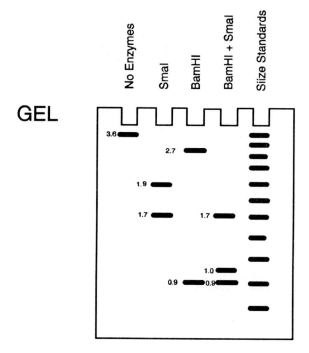

Figure 5-1
A **RESTRICTION MAP** is used to identify and locate specific restriction sites on a given piece of DNA. The size of a fragment is determined by running the restriction digest on an agarose gel. Fragments separate by size—the smaller ones move farther toward the bottom of the gel.

restriction fragments is near this length. Fortunately, they are not exactly this length, or they wouldn't be very useful.

The sequence of DNA recognized by a specific restriction endonuclease is often palindromic. A *palindrome* is something that reads the same way backward and forward (Fig. 5-2). The sequence of the bottom strand read in the 5' to 3' direction is the same as that of the top strand read in the 5' to 3' direction. The usual analogy for a verbal palindrome is a sentence that reads the same way backward and forward. "Madam, I'm Adam" is the usual example. It's not exactly the same way for DNA palindromes. The top strand does not read the same from the left as from the right; the top strand read from left to right is the same as the *bottom* strand read from right to left.

BamHI and other restriction endonucleases are dimeric enzymes that bind to a DNA palindrome and cut both strands at equivalent positions. The cut leaves two ends with complementary overhangs that will

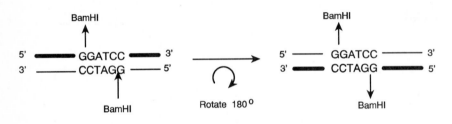

The BamHI site is palandromic - rotate it in the plane of the paper by 180° and the recognition sequence doesn't change.

Cutting with a restriction enzyme
generates two ends that are
complementary to each other

Figure 5-2
Useful **RESTRICTION ENZYMES** cleave DNA at symmetrical sites, leaving ends that are complementary.

hybridize to each other. The two ends can be rejoined later, or the fragment can be combined with other pieces of DNA cut with the same restriction endonuclease.

There can be a problem when using a single endonuclease to cut and rejoin different DNA fragments. The DNA fragments that result from cutting with a single restriction enzyme are the same at the two ends.[1] They can (and do) recombine with other pieces of DNA cut with the same restriction enzyme in either of two orientations—forward and backward. Since the DNA between the different restriction sites is not palindromic, the two orientations are not really equivalent, particularly if you're trying to make a protein by translating this region.

This problem can be solved by cutting the two pieces of DNA you want to join with two different restriction enzymes. This way the two ends of the DNA are not equivalent and the two cut pieces can be joined so that the DNA fragments can combine in only one orientation. This approach is very useful for joining different DNA fragments and inserting one specific piece of DNA into another specific piece of DNA. As we'll see a little later, putting inserts (translate as the piece of DNA you're interested in) into vectors (translate as something to carry your DNA around in) is essential to using recombinant-DNA techniques for sequencing, expressing, and mutating your protein (Fig. 5-3).

GELS AND ELECTROPHORESIS

These separate molecules by size—smaller ones move farther.

Gels are indispensable tools for the molecular biologist. Agarose or polyacrylamide can be formed into hydrophilic polymers that form hydrated gels in water. The gels are usually cast into thin, flat sheets between two plates of glass. The porous network in these gels retards the movement of macromolecules through them so that smaller molecules move faster. The size of the holes in the polymer can be changed by varying the amount of agarose or polyacrylamide in the gel. An electric

---

[1] Try rotating the DNA fragment by 180° in the plane of the paper (this means don't pick it up and flip it over—just turn the page upside down). You'll see that the ends look exactly the same as without the rotation. However, the middle, which is not palindromic, will be different.

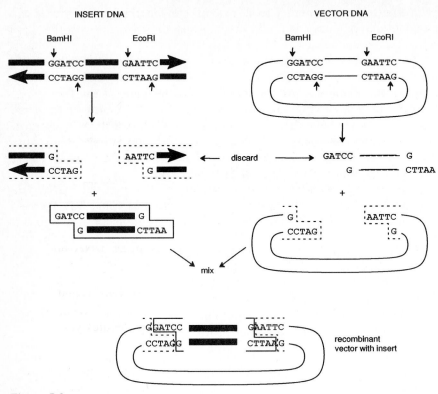

Figure 5-3
Generating a **RECOMBINANT-DNA** molecule using restriction enzymes to generate ends that can be joined in a specific fashion.

field applied across the ends of the gel causes the macromolecules to move. (DNA is negative and moves to the + electrode, which is at the bottom of the gel.) Molecules of the same size move the same distance, forming a band. Samples of DNA are applied to the top of the gel by putting them in slots (wells) formed during the casting operation. After electrophoresis, the molecules can be visualized by staining. A number of different stains can be used. Commonly, DNA is visualized by staining the gel with ethidium bromide, a dye that becomes intensely fluorescent when it intercalates into DNA. Radioactive nucleic acid fragments can be visualized by placing a piece of x-ray film against the gel. By comparing the distance a given band moves to the mobility of a series of standards of known size, the length of the DNA can be estimated.

BLOTTING

This means looking at specific molecules on gels even though there are many other molecules present that have the same size.

MOLECULE ON GEL	LABELED PROBE	NAME OF BLOT
DNA	DNA	Southern
RNA	DNA	Northern
Protein	Antibody	Western

The beauty of blotting techniques is that they let you see only what you're interested in. Take a whole gene's worth of DNA and make fragments with a restriction enzyme. Then separate these fragments by size on an agarose gel. Since there's lots of DNA in a genome, there will be lots of different DNA fragments of almost every size. Usual staining methods would show only a smear over the whole gel. What blotting techniques allow you to do is to detect only the molecules you're interested in.

After separating the molecules based on size, all the DNA fragments are transferred from the agarose gel to a piece of nitrocellulose paper.[2] The paper is actually placed against the gel, and the DNA molecules in the gel migrate from the gel to the paper, where they stick. The paper is then removed and heated to denature the DNA (it still sticks to the paper), and then the blot is cooled in the presence of a large excess of a radiolabeled, single-stranded DNA molecule (the probe) that contains the complement of the specific sequence that you want to detect. DNA fragments on the paper that contain sequences complementary to sequences in the probe will anneal to the radiolabeled probe. The excess probe is washed off, and the blot is placed against a piece of film. Only DNA fragments that have annealed to the probe will be radioactive, and a band will "light up" on the film everywhere there was a DNA molecule that contained sequences complementary to the probe. Conditions of hybridization (salt and temperature) can be changed to make the hybridization more selective (this is called *increased stringency*) so that the extent of sequence complementary between the probe and the DNA that is detected must be quite high.

[2] Special paper that actually reacts chemically with the DNA to cross-link it to the paper can also be used.

As long as the probe can find enough homology, it will stick (anneal) to DNA fragments on the blot that are longer or shorter than the probe itself. In the example shown in Fig. 5-4, the DNA fragment of interest (center lane) shows up as a single band. In this sample, there is only one size of DNA that has a sequence complementary to the probe sequence. In the digest of genomic DNA, two bands light up with this probe. In the genomic DNA, the probe sequence occurs in two different EcoRI fragments of different size. This could mean that there is sequence homology between two different genes (coding for two different proteins) or that an EcoRI restriction site is missing in one of the two copies of the gene present in the genome, reflecting a heterozygous gene pattern (in which the gene is different on each of the two diploid chromosomes).

These blotting techniques are known by the names of compass direc-

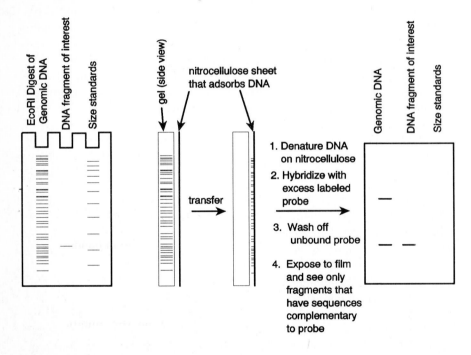

Figure 5-4
BLOTTING is a method to detect specific DNA (or RNA) fragments that contain sequences that are complementary to sequences in the labeled probe molecule. Only a few of the many DNA fragments on a gel will contain the sequence of interest, and only these will be seen (light up) on the blot. Specific proteins can also be visualized by blotting techniques using a specific antibody to detect a specific protein.

tions (Southern, Northern, Western). Since Southern is a person's name, there's no logic in how the different blots were named. Southern developed a blot in which DNA on the blot is detected by a labeled DNA probe. It was then fairly logical that the next technique developed, detecting RNA on the blot with a DNA probe, should be called a Northern blot. Then things got carried away with the Western, and now the Southwestern, and so on and so on.

If the gel separates DNA and the DNA is detected with a DNA probe, it is called a Southern blot. If RNA is separated on the gel and then detected by a DNA probe, it is a Northern. A Western uses specific antibodies to detect specific protein molecules on a blot of a protein gel. In the Western blot, the role of the DNA probe is filled by an antibody that recognizes a specific protein.

RESTRICTION FRAGMENT–LINKED POLYMORPHISM

RFLP is a Southern blot used to detect genetic disease.

For the diagnosis of genetic disease, some specific way of detecting a single mutation in DNA from the fetus must be used. The most obvious way to do this would be to use a restriction enzyme that cuts the wild-type sequence but does not cut the mutant sequence (or vice versa). A restriction site right at the site of the mutation would come in handy. If the fetal DNA has the normal sequence, the DNA will be cut and the restriction pattern will be identical to the wild type. If not, not. For many genetic diseases, the mutation does not conveniently occur right at a restriction site. However, in many cases, it just happens that the mutation that's being diagnosed is associated with another, nearby mutation that does alter some endonuclease cleavage site. This second site is closely linked genetically to the mutation that leads to the genetic disease. If the patient has this secondary restriction site, it's a good bet he or she has the mutation as well. The patterns that are observed when genomic DNA is digested with different endonucleases and the DNA is probed with a specific sequence can then be used to determine if a particular patient is homozygous or heterozygous for the specific mutation—a useful diagnostic tool.

CLONING

Cloning is manipulating a specific piece of DNA so that it can be used to generate multiple copies of itself or the RNA and protein that it encodes.

STEPS IN CLONING DNA:

1. Identify the DNA you want.
2. Put the DNA into a vector.
3. Change the sequence of the DNA (this is optional).
4. Put your DNA back into cells.
5. Grow the cells with your DNA/RNA/protein.

There are many different ways to clone a specific piece of DNA, but basically, they all involve (1) identifying and isolating the DNA you are interested in, (2) putting this DNA into something (a vector) to move it around from cell to cell, (3) altering the DNA sequence, (4) introducing this new DNA back into cells, and (5) growing the cells that have your DNA, RNA, or protein. You often do all this randomly to millions and millions of cells and then just select the few cells that got the piece of DNA you're interested in.

• **1. IDENTIFYING YOUR DNA:** There's lots of DNA out there, and finding just the right piece of DNA can be like finding a word in a dictionary that's arranged randomly.[3] The way you go about finding your DNA may depend on the reason for wanting the DNA in the first place.[4] The DNA you want will be contained in the genome of some cell. A frequent strategy is to take all the DNA in a specific cell, cut it into small fragments with restriction endonucleases, and put all these fragments into individual vectors (this is called a *genomic library*). A vector is a piece of DNA that makes it easy to capture other DNA fragments and move them around. Each individual vector will have only one piece of DNA inserted; however, the collection of vectors will contain much of the original cellular DNA. The same can be done with all the cell's mRNA, making

---

[3] You're right: "Arranged randomly" contradicts itself.

[4] The Mt. Everest rationale, "Because it's there," is not usually selective enough.

cDNA first using reverse transcriptase[5] (this is called a *cDNA library*). The DNA in a genomic library will contain introns, promoters, enhancers, and so forth; however, the DNA in a cDNA library will not contain introns or promoters, but it will contain a stretch of A's from the poly(A) tail of the mRNA. After introducing all this DNA into cells under conditions under which each cell will get only one of the DNA fragments in the library, the few cells that have your specific DNA will be identified.

Identification is easiest if your DNA confers some selective advantage to the cell (that is, if it expresses drug resistance or directs a function that is essential for cell survival under the conditions of your culture). Under selective conditions, only the cells with your DNA will survive. Killing cells (or, more mercifully, letting them die) that don't have the desired piece of DNA is called *selection*. A large number of cells (a million or so) can be spread on a culture plate, and only the ones that survive selection will continue to grow. These surviving colonies can be selected individually. If your DNA codes for a protein and you have antibody to the protein or the protein has an activity that is not present in the host cell, the cells with your DNA can be detected by looking for the cells that make the protein or have the activity. Finding the cells with your DNA by detecting the DNA directly with a Southern blot, or by detecting the protein or RNA product of the gene, is called *screening*.

It's also possible to select your DNA before you put it in the vector. If you know the sequence (or even part of it), DNA pieces (from genomic DNA or cDNA) with this sequence can be purified on a gel and identified by hybridization to an oligonucleotide using a Southern blot. Alternatively, if you know the sequence of the ends of your DNA, you can amplify it specifically by the polymerase chain reaction. There are lots of clever ways to find your DNA.

• 2. PUTTING YOUR DNA INTO A VECTOR: Vectors are specialized pieces of DNA used to move other pieces of DNA around. Modern vectors are usually either bacterial plasmids or viral genomes. The act of isolating your DNA in the first place usually involves putting it into a vector and then selecting the vector that has your DNA in it. DNA pieces (called *inserts* when they are placed in a vector) are usually placed in vectors using restriction endonucleases. The vector is cut with two restriction enzymes of different specificity (Fig. 5-3). This removes a chunk of the vector DNA and leaves two different ends. You then cut your DNA

---

[5] Reverse transcriptase is an enzyme isolated from viruses that contain a genome that is RNA. This viral enzyme makes DNA using RNA as a template.

with the same two enzymes so that it will have the same complementary ends. Restriction enzymes that don't cut in an essential part of the vector or insert must be used. You can also make suitable cloning sites by cutting with just one restriction enzyme; however, because of the palindromic nature of restriction enzyme specificity, the ends of the piece of DNA will be the same. The DNA can then go into the vector in either one of two orientations. Sometimes this matters and sometimes it doesn't. If you want RNA or protein expressed from your DNA, direction will matter if the promoter site is provided by the vector. After mixing your cut DNA with the cut vector under conditions under which the ends will anneal, DNA ligase (and ATP) is added to join the strands with a covalent bond.

Vectors are often designed to contain a drug-resistance marker to aid in the selection of cells that have incorporated your vector (not all cells do). They can also have a variety of other goodies depending on the type of vector. An expression vector is used to express RNA or protein from the DNA, and these vectors usually contain a good promoter region and some way to turn the promoter on and off. Many expression vectors have been engineered to contain a convenient set of unique restriction sites (termed a *polylinker*) near the promoter to make it easy to put your insert in the right place. Sequencing vectors, designed to make it easy to sequence your DNA, usually have a defined site for the sequencing primer to bind that is adjacent to a polylinker region.

• 3. CHANGING THE SEQUENCE OF YOUR DNA.   The sequence of the DNA can be changed in lots of ways. Large chunks can be deleted or added (deletion or insertion mutagenesis) by mixing and matching endonuclease fragments. Sequences of DNA from one gene can be combined with sequences from another gene (chimeric DNA—named for the Chimera, a mythological beast with the head of a lion, the tail of a serpent, and the body of a goat). If protein product is going to be made from the mutant DNA, care must be taken to preserve the reading frame. Deleting or inserting a number of bases that is not divisible by 3 will cause a shift in the reading of the triplet codons and a jumbling of the protein sequence. Individual nucleotides can be changed at any specific site by the use of site-directed mutagenesis. The reason for changing the DNA sequence is to change the function of the DNA itself or its RNA or protein product.

• 4. PUTTING YOUR DNA BACK INTO CELLS.   Vectors can be isolated and then added back to cells. DNA can be introduced into cells in a variety of ways: by infection with a virus containing your DNA, by poking holes in the cells with specific salt solutions, by precipitating the DNA with calcium phosphate and having cells take up the precipitate, by blowing holes in the cells with an electric discharge and allowing pieces of DNA to enter the cells through the holes (*electroporation*), or by directly microinjecting the DNA with a very small glass capillary.

Not all cells that are exposed to your vector will take it up. That's where selection is helpful. You just kill all the cells you're not interested in.

SEQUENCING

Sequencing is determining the sequential order of DNA bases in a given piece of DNA.

Sequencing DNA is relatively easy these days, at least for small pieces (a few thousand nucleotides). In the Sanger dideoxynucleotide method, a specific primer is used that is complementary to one of the two DNA strands you want to sequence. The primer can be a vector sequence so that you can sequence any piece of DNA cloned into the vector. The primer is a synthetic oligonucleotide that is radiolabeled (or fluorescently labeled) so that you can see all new DNA molecules that have the primer attached to the 5' end. Alternatively one of the deoxynucleotides used in the DNA synthesis can be labeled. After denaturing the double-stranded DNA that you want to sequence and annealing the primer, the DNA is elongated from the primer (in the 5' to 3' direction) using DNA polymerase. The reaction is run for a short time with all four deoxynucleotides. There will be pieces of DNA that are at all stages of the replication process—the newly synthesized DNA will be of all different lengths. The reaction is then stopped by adding it to four separate tubes, each of which contains a different 2', 3'-dideoxynucleotide. When a dideoxynucleotide is incorporated by the polymerase, the elongation stops (there's no 3'-hydroxyl group on the dideoxynucleotide). The trick is that only one of the four dideoxynucleotides will stop the reaction at any given point in the random mixture of newly synthesized DNA. The synthesized DNA is then run on a high-resolution acrylamide gel that can separate DNA molecules that differ in length by one nucleotide. Four lanes are run, one for each type of dideoxynucleotide used to stop the reaction. A ladder of bands will be seen. The shorter bands, at the bottom of the gel, will correspond to termination nearest the primer (near the 5' end). The sequence is then read from the bottom (5' end) to the top (3' end) of the gel by noting which dideoxynucleotide stopped the reaction at that length (that is, simply which one of the four lanes has a band in it at that length) (Fig. 5-5).

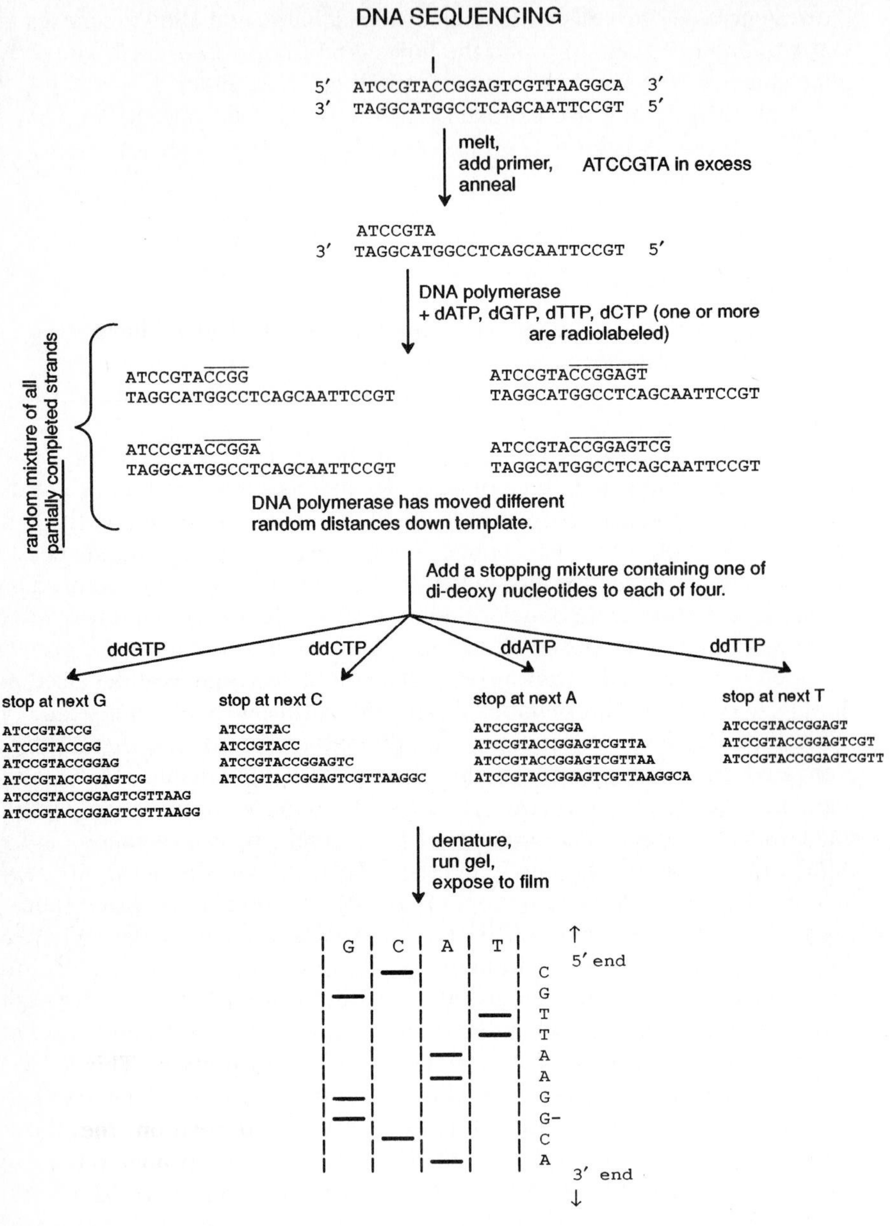

DNA SEQUENCING

|

5′ ATCCGTACCGGAGTCGTTAAGGCA 3′
3′ TAGGCATGGCCTCAGCAATTCCGT 5′

↓ melt,
add primer,   ATCCGTA in excess
anneal

ATCCGTA
3′ TAGGCATGGCCTCAGCAATTCCGT 5′

↓ DNA polymerase
+ dATP, dGTP, dTTP, dCTP (one or more
are radiolabeled)

random mixture of all partially completed strands {

ATCCGTA̅C̅C̅G̅G̅   ATCCGTA̅C̅C̅G̅G̅A̅G̅T̅
TAGGCATGGCCTCAGCAATTCCGT   TAGGCATGGCCTCAGCAATTCCGT

ATCCGTA̅C̅C̅G̅G̅A̅   ATCCGTA̅C̅C̅G̅G̅A̅G̅T̅C̅G̅
TAGGCATGGCCTCAGCAATTCCGT   TAGGCATGGCCTCAGCAATTCCGT

DNA polymerase has moved different
random distances down template.

↓ Add a stopping mixture containing one of
di-deoxy nucleotides to each of four.

ddGTP ←   ddCTP ↙   ddATP ↘   ddTTP →

stop at next G	stop at next C	stop at next A	stop at next T
ATCCGTACCG	ATCCGTAC	ATCCGTACCGGA	ATCCGTACCGGAGT
ATCCGTACCGG	ATCCGTACC	ATCCGTACCGGAGTCGTTA	ATCCGTACCGGAGTCGT
ATCCGTACCGGAG	ATCCGTACCGGAGTC	ATCCGTACCGGAGTCGTTAA	ATCCGTACCGGAGTCGTT
ATCCGTACCGGAGTCG	ATCCGTACCGGAGTCGTTAAGGC	ATCCGTACCGGAGTCGTTAAGGCA	
ATCCGTACCGGAGTCGTTAAG			
ATCCGTACCGGAGTCGTTAAGG			

↓ denature,
run gel,
expose to film

Figure 5-5   DNA Sequencing

MUTAGENESIS

Making a mutant DNA
Deletion: Deletes a hunk of DNA
Insertion: Inserts a hunk of DNA
Site-directed: Modifies a specific nucleotide
Random: Introduces random changes in the DNA

Mutagenesis is used to alter the DNA structure (sequence) in a known way by either deleting nucleotides, inserting nucleotides, or changing a single nucleotide at a defined location. Random mutagenesis of DNA may be performed over the whole piece of DNA by exposing the DNA to chemicals (mutagens) that react with the DNA and change the specificity for base pairing or by using oligonucleotides that contain random, deliberate mistakes in the sequence. Mutants are then selected or screened for changes in function of the protein or RNA product. This technique allows you to define specific amino acids that are essential to the function of the protein and to determmine which amino acids can be replaced by which other amino acids and still conserve function.

In deletion or insertion mutagenesis, restriction enzymes are used to generate DNA with a specific fragment missing or with another piece of DNA inserted. This change in the DNA sequence can then be used to produce RNA or protein containing a deletion or insertion of amino acids in the protein. This is useful in determining the gross features of the gene structure that are necessary to preserve a functional gene or to express a functional protein product. For example, a signal sequence that directs the synthesized protein to the mitochondrial matrix can be placed in the sequence of a protein that is normally cytosolic. This mutant protein will be expressed as a mitochondrial matrix protein.

With site-directed mutagenesis, a change in the DNA sequence can be introduced at any specific site. An oligonucleotide is annealed to a single-strand copy of the DNA that you want to mutate. This oligo-nucleotide contains the correct (complementary) base at every position except the one you want to change. At the mutated position, there is a mismatch. After the oligonucleotide is annealed at the proper position, the DNA is fully replicated using DNA polymerase and then sealed with ligase. When the vector is introduced into the host cell and replicated during cell division, some of the progeny cells will get DNA that has used the mutant strand as the template for DNA replication. There are clever ways to increase your chances of getting only cells containing the mutant DNA. These involve selectively destroying the wild-type (nonmutated) strand. Site-directed mutagenesis is used to change single amino acids in proteins or single bases in RNA or DNA. The technique has been very useful in determining the function of a specific amino acid residue in enzyme catalysis, binding of a ligand, or stabilizing a protein. It has also been possible to selectively change the activity and specificity of some enzymes using this technique (Fig. 5-6).

POLYMERASE CHAIN REACTION

PCR amplifies DNA sequences that lie between specific 5′ and 3′ sequences.

This is a technique for amplifying a specific segment of DNA. Oligonucleotide primers are synthesized that are complementary to one strand at the 5′ end of your DNA and complementary to the opposite strand at the 3′ end. After the DNA is denatured and the oligonucleotide primers (in excess) are annealed, the DNA is elongated using DNA polymerase and deoxynucleotides. A new double-stranded DNA molecule will be generated starting from each primer. DNA sequences behind (to the 5′ side of) the primer will not be replicated. The DNA is then heated to denature it and reannealed to the primer again. Another round of replication is performed. This cycle is repeated over and over, with a twofold increase in the amount of DNA on each cycle. Because two primers are used, only the sequence *between* the two primers will be amplified. Since the cycle is carried out multiple times with a twofold increase in the amount of DNA each time, a geometric amplification results (10 cycles would result in a 2^{10} increase in the DNA concentration). The limitation on the amount of DNA that is produced is the amount of primer and deoxynucleotides added. The cleverness of this technique is extended by using a heat-stable DNA polymerase that is not inactivated by the temperatures needed to denature the DNA. Multiple cycles can be

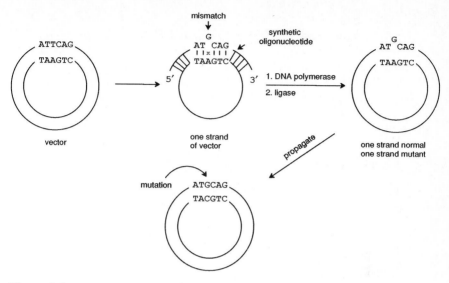

Figure 5-6
SITE-DIRECTED MUTAGENESIS can be used to change one or more base pairs in the DNA resulting in a change in the amino acid that appears in the protein produced from this DNA.

performed simply by heating (denaturing) and cooling (renaturing and polymerizing) a tube containing the DNA, the primers, deoxynucleotide triphosphates, and the DNA polymerase. Because of the extreme amount of amplification, PCR can be used to amplify sequences from very small amounts of DNA. New restriction sites can be easily generated by including them in the 5' end of the oligonucleotide primer even though they are not present in the original DNA. As long as the primer is still long enough to hybridize to the DNA through complementary sequences, the dangling 5' ends containing the restriction site sequence will be amplified in the next round. PCR can also be used to remove inserts from vectors and to introduce site-specific mutants (Fig. 5-7).

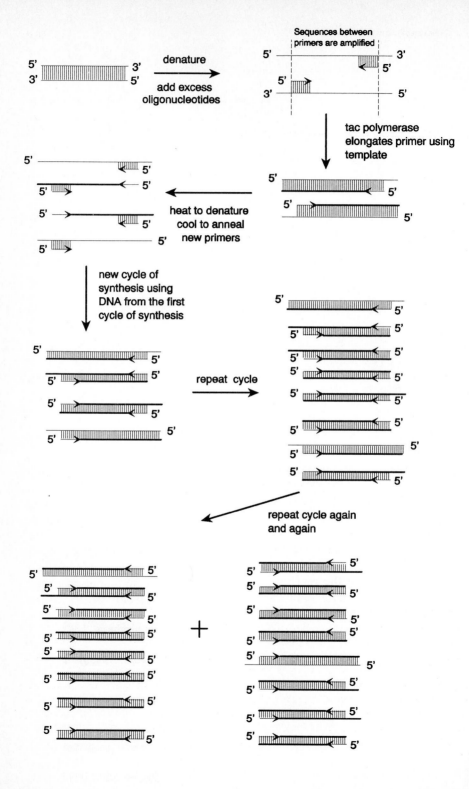

← **Figure 5-7   The Polymerage Chain Reaction**
PCR is used to amplify (synthesize) specific DNA sequences that lie between a 5′ primer and a 3′ primer. The primers are annealed to the appropriate DNA strand and are lengthened (5′ to 3′) by adding deoxynucleotides, using DNA polymerase and the longer DNA strand as a template. The newly synthesized DNA is dematured by heating, cooled to allow more primer to anneal to the newly synthesized strands, and the cycle of synthesis, melting, and annealing new primer is repeated over and over. Each cycle increases the amount of DNA by twofold. Note that with increasing numbers of cycles the sequences between the two primers are amplified more than sequences outside the primers.

ENZYME MECHANISM

·

Active Site

Transition State

Catalysis

Lock and Key

Induced Fit

Nonproductive Binding

Entropy

Strain and Distortion

Transition-State Stabilization

Transition-State Analogs

Chemical Catalysis

· · · · · · · · · · · ·

Enzymes do two important things: they recognize very specific substrates, and they perform specific chemical reactions on them at fantastic speeds. The way they accomplish all this can be described by a number of different models, each one of which accounts for some of the behavior that enzymes exhibit. Most enzymes make use of all these different mechanisms of specificity and/or catalysis. In the real world, some or all of these factors go into making a given enzyme work with exquisite specificity and blinding speed.

ACTIVE SITE

The active site is a specialized region of the protein where the enzyme interacts with the substrate.

The active site of an enzyme is generally a pocket or cleft that is specialized to recognize specific substrates and catalyze chemical transformations. It is formed in the three-dimensional structure by a collection of different amino acids (active-site residues) that may or may not be directly connected in the primary sequence. The interactions between the active site and the substrate occur via the same forces that stabilize protein structure: hydrophobic interactions, electrostatic interactions (charge–charge), hydrogen bonding, and van der Waals interactions. Enzyme active sites do not simply bind substrates; they also provide catalytic groups to facilitate the chemistry and provide specific interactions that stabilize the formation of the transition state for the chemical reaction.

TRANSITION STATE

The transition state is the highest-energy arrangement of atoms during a chemical reaction.

During a chemical reaction, the structure of the substrate changes into the structure of the product. Somewhere in between, some bonds are partly broken; others are partly formed. The transition state is the highest-energy arrangement of atoms that is intermediate in structure between the structure of the reactants and the structure of the products. Figure 6-1 is called a *reaction coordinate diagram*. It shows the free energy of the reactants, transition state, and product. The free-energy difference between the product and reactant is the free-energy change for the overall reaction and is related to the equilibrium constants [$\Delta G = -RT\ln(K_{eq})$].[1] The free-energy change between the products and reactants tells you how favorable the reaction is thermodynamically. It doesn't tell you anything about how fast.

Reactions don't all occur with the same rate. Some energy must be put into the reactants before they can be converted to products. This activation energy provides a barrier to the reaction—the higher the bar-

[1] There is a more complete description of thermodynamics in Chap. 22.

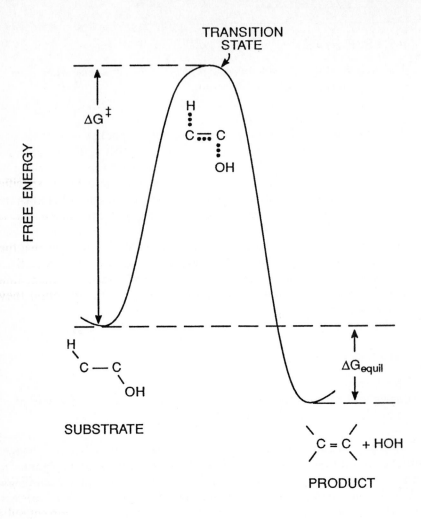

Figure 6-1
FREE-ENERGY CHANGES occur during a chemical reaction. The least stable arrangement of the atoms during the reaction is called the transition state, and it occurs at an energy maximum. How fast the reaction happens is determined by the free energy of activation, the free-energy difference between the substrate and the transition state. The larger the free energy of activation, the slower the reaction.

rier, the slower the reaction. The difference in free energy between the transition state and the reactant(s) is called the *free energy of activation*.

CATALYSIS

The reaction happens at a faster rate.
The catalyst is regenerated.

Enzymes do chemistry. Their role is to make and break specific chemical bonds of the substrates at a faster rate and to do it without being consumed in the process. At the end of each catalytic cycle, the enzyme is free to begin again with a new substrate molecule.

Since catalysis is simply making a reaction go faster, it follows that the activation energy of a catalyzed (faster) reaction is lower than the activation energy of an uncatalyzed reaction. It's possible to say, then, that enzymes work by lowering the activation energy of the reaction they catalyze. This is the same as saying that enzymes work because they work. The question is how they lower the activation energy.

LOCK AND KEY

Specificity model—the correct substrate fits into the active site of the enzyme like a key into a lock. Only the right key fits.

This is the oldest model for how an enzyme works. It makes a nice, easy picture that describes enzyme specificity. Only if the key fits will the lock be opened. It accounts for why the enzyme only works on certain substrates, but it does not tell us why the reaction of the correct substrates happens so fast. It doesn't tell us the mechanism of the lock. A problem arises because the structure of the substrate changes as it is converted to product. So what is the enzyme complementary *to*—the substrate, the product, or what? The answer is often the transition state (Fig. 6-2).

INDUCED FIT

The binding of the correct substrate triggers a change in the structure of the enzyme that brings catalytic groups into exactly the right position to facilitate the reaction.

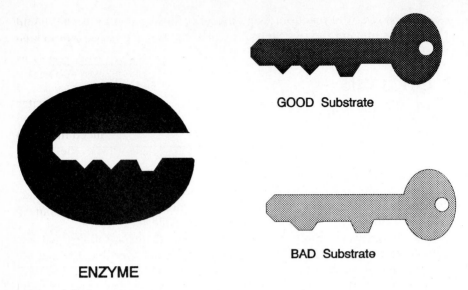

GOOD Substrate

BAD Substrate

ENZYME

Figure 6-2
The **LOCK AND KEY** model for enzyme specificity uses complementarity between the enzyme active site (the lock) and the substrate (the key). Simply, the substrate must fit correctly into the active site—it must be the right size and shape, have charges in the correct place, have the right hydrogen-bond donors and acceptors, and have just the right hydrophobic patches.

In the induced-fit model, the structure of the enzyme is different depending on whether the substrate is bound or not. The enzyme changes shape (undergoes a conformation change) on binding the substrate. This conformation change converts the enzyme into a new structure in which the substrate and catalytic groups on the enzyme are properly arranged to accelerate the reaction. "Bad" substrates cannot cause this conformation change. For example, the enzyme hexokinase catalyzes the transfer of phosphate from ATP to the 6-hydroxyl group of glucose.

$$\text{Glucose—OH} + \text{ATP} \longrightarrow \text{Glucose—O—P} + \text{ADP}$$

Chemically, glucose—OH is very similar in reactivity to water; it's just got some other structural parts that make it look more complicated.

$$\text{H—OH} + \text{ATP} \longrightarrow \text{H—O—P} + \text{ADP}$$

Although water and glucose are chemically similar, hexokinase catalyzes the transfer of phosphate to glucose about 10^5 times faster than it catalyzes the transfer of phosphate to water.[2] The induced-fit model would argue that the fancy part of the glucose molecule is necessary to induce the enzyme to change its conformation and become an efficient catalyst. Even though the fancy part of the glucose molecule is not directly involved in the chemical reaction, it participates in the enzyme-catalyzed reaction by inducing a change in the structure of the enzyme. Since water doesn't have this extra appendage, it can't cause the conformation change and is, therefore, a poor substrate for this enzyme. The induced fit-model would say that in the unreactive conformation of the enzyme, the ATP is 10^5 times less reactive than when the enzyme is in the reactive conformation (Fig. 6-3).

What the induced-fit model is good at explaining is why bad substrates are bad, but like the lock and key model, it too fails to tell us exactly why good substrates are good. What is it about the "proper" arrangement that makes the chemistry fast?

NONPRODUCTIVE BINDING

Poor substrates bind to the enzyme in a large number of different ways, only one of which is correct. Good substrates bind only in the proper way.

Again this model tells us why poor substrates don't work well. Poor substrates bind more often to the enzyme in the wrong orientation than in the right orientation. Since poor substrates bind in the wrong orientation, the catalytic groups and specific interactions that would accelerate the reaction of the correct substrate come into play in only a very small number of the interactions between the enzyme and a bad substrate. In contrast to the induced-fit model, this model does not require a change in the conformation of the enzyme (Fig. 6-4). In the hexokinase reaction discussed above, the nonproductive binding model would say that only 1 out of 10^5 water molecules binds to the enzyme in a productive fashion but all the glucose binds in a productive orientation.

ENTROPY

Organizing a reaction at the active site of an enzyme makes it go faster.

---

[2] Luckily, water is not a good substrate for hexokinase. Otherwise the ATP hydrolysis would burn up a good part of the ATP we ate so hard for.

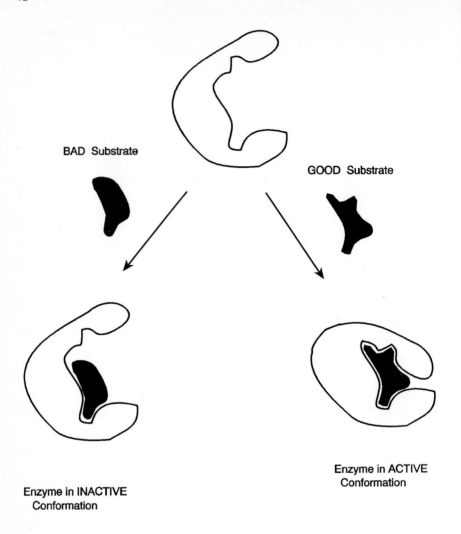

Figure 6-3
The **INDUCED-FIT** model for enzyme specificity says that good substrates must be able to cause the enzyme to change shape (conformation) so that catalytic and functional groups on the enzyme are brought into just the right place to catalyze the reaction. Bad substrates are bad because they aren't able to make the specific interactions that cause the conformation change, and the enzyme stays in its inactive conformation,.

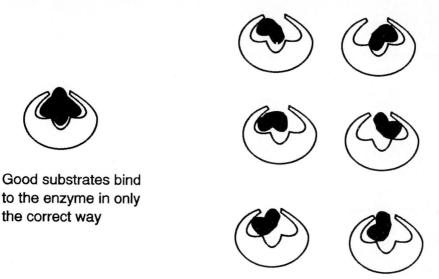

Good substrates bind
to the enzyme in only
the correct way

Bad substrates bind to the
enzyme in a large number
of incorrect, nonproductive
ways

Figure 6-4
The **NONPRODUCTIVE BINDING** model suggests that while good substrates
bind in only one, correct way, bad substrates usually bind to the enzyme incor-
rectly and cannot react.

When molecules react, particularly when it's the reaction between
two different molecules or even when it's a reaction between two parts of
the same molecule, they must become more organized. The reason is that
the two reacting atoms must approach each other in space. Just finding
the appropriate partner is often a tough part of the reaction (biochemistry
mimics life once more). Part of the free-energy barrier to a chemical
reaction is overcoming unfavorable entropy changes that must accompany the formation of the transition state. By binding two substrates at
the same active site, the enzyme organizes the reacting centers. This
reduces the amount of further organization that must occur to reach the
transition state for the reaction, making the free energy of activation
lower and the reaction faster. Of course, the enzyme must find each of the
substrates and organize them at the active site—this is entropically unfavorable too. However, the price paid for organizing the substrates can
be taken out of the binding free energy. (In English, the last sentence

means that organizing the substrates at the active site makes the sub-strates bind to the enzyme less tightly than they would if they didn't have to become organized. This is not a big deal since we can just increase the concentration of the substrate to what is necessary to make it bind.) Entropy and disorder make good dinner conversation topics—right up there with supply-side economics and the national debt.

STRAIN AND DISTORTION

The binding of the substrate results in a distortion of the substrate (pulling or pushing on a bond) in a way that makes the chemical reaction easier.

In this model, the binding of the substrate to the enzyme strains specific chemical bonds, making the subsequent chemical reaction easier. If a bond has to be broken, the enzyme grabs onto both sides of the bond and pulls. If a bond has to be formed, the enzyme grabs onto both sides and pushes. By this model, the enzyme must be designed to apply the strain in the right direction—the direction that will help convert the reactant to the transition state (Fig. 6-5).

TRANSITION-STATE STABILIZATION

Enzymes recognize and stabilize atomic features that are present in the transition state for the catalyzed reaction but that are not present in the substrate or product. The enzyme is more complementary to the transition state than to the reactants of products.

A common theme of all of the above mechanisms of catalysis is that the enzyme does something to assist the reaction in reaching the transi-tion state. The structure of the transition state for a chemical reaction is slightly different from the structure of either the reactants or products. Some chemical bonds are at different angles and lengths, and charges are distributed differently. The enzyme can stabilize those features that occur just in the transition state by providing groups at just the right location and orientation to interact with the transition state and not with the substrates or products. In other words, we can view the enzyme active site as being complementary to the transition state.

How can an enzyme specifically recognize the transition state? Let's

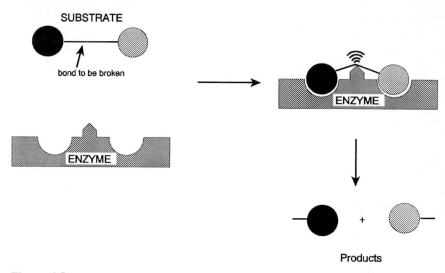

Figure 6-5
The **STRAIN AND DISTORTION** model for catalysis involves pushing, pulling, or twisting a bond that is to be made or broken during the reaction. Parts of the substrate not involved directly in the chemical reaction are required to hold the substrate on the enzyme in the distorted form. The distortion and strain make it easier to reach the transition state.

pick a simple chemical reaction like the addition of an alcohol to a phosphate ester (Fig. 6-6). In this reaction, negative charge develops on the phosphate oxygen, and the bond angles to phosphorous change during the reaction. The enzyme can be designed so that there are hydrogen-bond acceptors or a positive charge located at exactly the right position to interact with the charge that has to develop in the transition state. This favorable interaction will be present only when the substrate is in the transition state structure—otherwise it won't contribute favorably to binding. There are two equivalent ways of saying what this additional favorable interaction with the transition state actually does. First, we can say that the interaction stabilizes (lowers the free energy of) the transition state more than it does the ground state (substrates and products). Second, we can say that the enzyme binds the transition state more tightly than the ground state(s).

If an enzyme binds the transition state for a chemical reaction more tightly than the ground state (substrate), the reaction must go faster. Let's see if I can convince you of that using a reaction coordinate diagram (Fig. 6-7). The reaction coordinate diagram is shown for any old generic chemical reaction (solid line). When the substrate interacts with the enzyme,

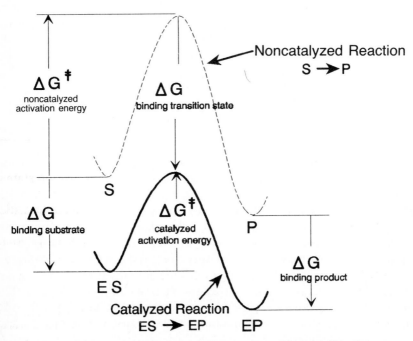

Figure 6-6
The **STRUCTURE OF THE TRANSITION STATE** is different from that of the substrate with respect to charge and shape. Because it looks different, the enzyme can recognize specific features of the transition state and stabilize them. This makes it easier to reach the transition state and makes the reaction faster.

Figure 6-7   Transition-State Binding Is Tighter Than Substrate Binding
The ΔG for binding the substrate and the transition state is shown as a difference between the energies of the ES complex and E + S. The ΔG for binding the transition state is shown as a difference between the energies of the E · TS complex and E + TS. If the transition state binds tighter (bigger ΔG) than the substrate, the enzyme-catalyzed reaction must have a lower activation energy.

there is a free-energy change.[3] The tighter (more favorable) the interaction, the more stable the ES complex, and the lower the free energy. If the enzyme binds the transition state more tightly than the substrate or product, the free-energy *change* on binding the transition state will be more negative (more favorable) than the free energy change for binding the substrate. When the free energies of the enzyme-bound substrates, products, and transition states are compared (dotted line), you'll see that the free energy of activation for the reaction has decreased and the reaction must be faster. The rather magic conclusion is that any favorable and specific interaction that the enzyme makes with the transition state makes the reaction go faster.

There are other parts of the substrate, not involved in the reaction, that don't change structure during the reaction—like the R groups. These groups and the interactions the enzyme makes with them are also important. It is the interaction of the enzyme with groups that are the same in the ground state and transition state that allows the enzyme to hold onto the substrate, transition state, and product as the reaction proceeds on the enzyme.

TRANSITION-STATE ANALOG

This is a molecule that is designed to look like the transition state for a specific chemical reaction.

If an enzyme recognizes (is complementary to) the transition state of the reaction, it should be possible to construct molecules that bind very tightly to the enzyme by making a molecule that looks like the transition state. Transition states themselves can't be isolated—after all, they're not a stable arrangement of atoms, and some bonds are only partially formed

---

[3] For a binding reaction we can pick whether we show the reaction as favorable or unfavorable by picking the substrate concentration we use. Association constants have concentration units (M^{-1}). The equilibrium position of the reaction (how much ES is present) depends on what concentration we pick for the substrate. At a concentration of the substrate that is much less than the dissociation constant for the interaction, most of the enzyme will not have substrate bound, the ratio[ES]/[E] will be small, and the apparent equilibrium constant will also be small. This all means that at a substrate concentration much less than the dissociation constant, the binding of substrate is unfavorable. At substrate concentrations higher than the dissociation constant, most of the enzyme will have substrate bound and the reaction will be shown as being favorable (downhill). (See also the discussion of saturation behavior in Chap. 7.)

or broken. But for some enzymes, analogs can be synthesized that are stable but still have some of the structural characteristics of the transition state (Fig. 6-8).

Proteolytic enzymes catalyze the hydrolysis of peptide bonds by forming a covalent intermediate with the substrate. The formation of this intermediate involves the addition of a group from the enzyme (usually a Ser–OH or Cys–SH) to the carbonyl of the amide bond. In the transition state for this reaction (actually just part of the reaction), the carbon atom goes from a planar configuration to a tetrahedral arrangement, and the carbonyl oxygen develops negative charge. Phosphonate analogs of the peptide bond are tetrahedral and have a negatively charged oxygen. They are excellent inhibitors of proteolytic enzymes, because they look more like the transition state for the reaction than like the substrate.

CHEMICAL CATALYSIS

The amino acid side chains and enzyme cofactors provide functional groups that are used to make the reaction go faster by providing new pathways and by making existing pathways faster.

Many chemical reactions can be made to occur faster by the use of appropriately placed catalytic groups. Enzymes, because of their three-dimensional structure, are great at putting just the right group in the right place at the right time. Take the simple reaction of the addition of water to a carbonyl group (Fig. 6-9). We can talk about two factors with this one reaction. The carbonyl group is reactive toward water because the carbonyl group is polarized—the electrons of the $C=O$ are not shared equally between the carbon and the oxygen. The carbon atom has fewer of them (because oxygen is more electronegative). As the water attacks the carbonyl oxygen, the electrons in the bond being broken ($C=O$) shift to oxygen, giving it a formal negative charge. Putting a positively charged group near the oxygen of the carbonyl group polarizes the carbonyl group and makes the carbonyl more reactive by helping stabilize the development of negative charge on oxygen as the chemical reaction proceeds (*electrostatic catalysis*).

Now let's look at what we can do with the water. Because it has more negative charge (a higher electron density), ^-OH is more reactive than HOH. By providing an appropriately placed base to at least partially remove one of the protons from the attacking water molecule, we can increase the reactivity of this water and make the reaction go faster. This is known as *acid–base catalysis* and is widely used by enzymes to help facilitate the transfer of protons during chemical reactions.

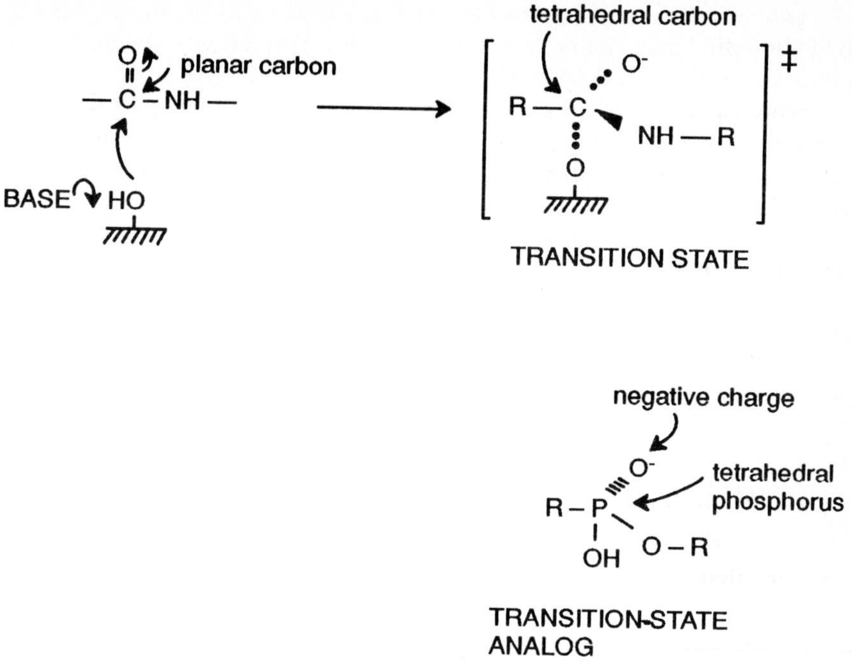
planar carbon
BASE
HO
tetrahedral carbon
R
C
O
NH — R
TRANSITION STATE
negative charge
tetrahedral phosphorus
R – P
OH
O – R
TRANSITION-STATE ANALOG

Figure 6-8
TRANSITION-STATE ANALOGS are stable molecules that are designed to look more like the transition state than like the substrate or product. Transition-state analogs usually bind to the enzyme they're designed to inhibit much more tightly (by 1000-fold or more) than the substrate does.

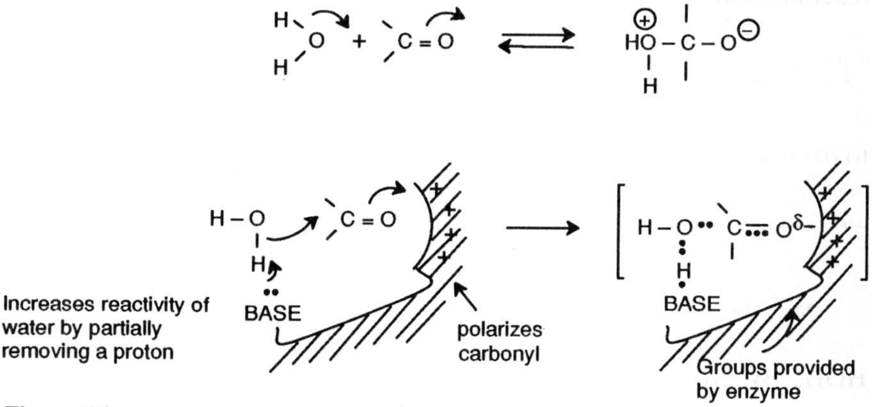
C = O
HO – C – O
Increases reactivity of water by partially removing a proton
BASE
polarizes carbonyl
BASE
Groups provided by enzyme

Figure 6-9
An enzyme can **STABILIZE THE TRANSITION STATE** by providing specific interactions with developing charge and shape features that are present only in the transition state. These interactions are not available to the uncatalyzed reaction.

Another alternative is for the enzyme to actually form a covalent bond between the enzyme and the substrate. This direct, covalent participation of the enzyme in the chemical reaction is termed *covalent catalysis*. The enzyme uses one of its functional groups to react with the substrate. This enzyme–substrate bond must form faster and the intermediates must be reasonably reactive if this kind of catalysis is going to give a rate acceleration.

· C H A P T E R · 7 ·

ENZYME KINETICS

·

S, P, and E (Substrate, Product, Enzyme)

Amounts and Concentrations

Active Site

Assay

Velocity

Initial Velocity

Mechanism

Little k's

Michaelis-Menten Equation

V_{max}

k_{cat}

K_m

Special Points

k_{cat}/K_m

Rate Accelerations

Steady-State Approximation

Transformations and Graphs

Inhibition

Allosterism and Cooperativity

The Monod-Wyman-Changeaux Model

· · · · · · · · · · · ·

Kinetics seems scary, but understanding just a few things spells relief. Two problems with kinetics are the screwy (and often unexplained) units and the concepts of *rate* and *rate constant*. And you can't ignore enzyme kinetics; it forms the foundation of metabolic regulation, provides a diagnostic measure of tissue damage, and lies at the heart of drug design and therapy.

S, P, and E (SUBSTRATE, PRODUCT, ENZYME)

Enzyme (E) converts substrate(s) (S) to product(s) (P) and accelerates the rate.

The most well studied enzyme catalyzes the reaction $S \rightleftharpoons P$. The kinetic question is how time influences the amount of S and P present at any time. In the absence of enzyme, the conversion of S to P is slow and uncontrolled. In the presence of a specific enzyme (S-to-Pase[1]), S is converted swiftly and specifically to product. S-to-Pase is specific; it will not convert A to B or X to Y. Enzymes also provide a rate acceleration. If you compare the rate of a chemical reaction in solution with the rate of the same reaction with the reactants bound to the enzyme, the enzyme reaction will occur up to 10^{14} times faster.

AMOUNTS AND CONCENTRATIONS

AMOUNT	CONCENTRATION
A quantity	*A quantity/volume*
mg, mole	M (mol/L), μM
g, μmol	mM, mg/mL
units	units/mL

Quantities such as milligrams (mg), micromoles (μmol), and units refer to amounts. Concentration is the amount per volume, so that molar (M), micromolar (μM), milligrams per milliliter (mg/ml), and units per milliliter (units/ml) are concentrations. A unit is the amount of enzyme that will catalyze the conversion of 1 μmol of substrate to product in 1 min under a given set of conditions.

[1] Enzymes are named by a systematic set of rules that nobody follows. The only given is that enzyme names end in -ase and may have something in them that may say something about the type of reaction they catalyze—like chymotrypsin, pepsin, enterokinase (all proteases).

The concentrations of substrate and product are invariably in molar units (M; this includes mM, μM, etc.), but enzyme concentrations may be given in molar (M), milligrams per milliliter (mg/mL), or units/mL. The amount of enzyme you have can be expressed in molecules, milligrams, nanomoles (nmol), or units. A *unit* of enzyme is the amount of enzyme that will catalyze the formation of 1 μmol of product per minute under specifically defined conditions. A unit is an amount, not a concentration.

Units of enzyme can be converted to milligrams of enzyme if you know a conversion factor called the *specific activity*. Specific activity is the amount of enzyme activity per milligram of protein (micromoles of product formed per minute per milligram of protein, or units per milligram). For a given pure enzyme under a defined set of conditions, the specific activity is a constant; however, different enzymes have different specific activities. To convert units of enzyme to milligrams of enzyme, divide by the specific activity: units/(units/milligram) = milligrams. Specific activity is often used as a rough criterion of purity, since in crude mixtures very few of the milligrams of protein will actually be the enzyme of interest. There may be a large number of units of activity, but there will also be a large amount of protein, most of which is not the enzyme. As the enzyme becomes more pure, you'll get the same units but with less protein, and the specific activity will increase. When the protein is pure, the specific activity will reach a constant value.

Enzyme concentrations in milligrams per milliliter can be converted to molar units by dividing by the molecular weight (in mg/mmol).[2]

$$\frac{\text{mg/mL}}{\text{mg/mmol}} = \frac{\text{mmol}}{\text{mL}} = M$$

ACTIVE SITE

The active site is the special place, cavity, crevice, chasm, cleft, or hole that binds and then magically transforms the substrate to product. The kinetic behavior of enzymes is a direct consequence of the protein's having a limited number (often 1) of specific active sites.

Most of enzyme kinetics (and mechanism) revolves around the active site. As we'll see later, saturation kinetics is one of the direct consequences of an active site.

---

[2] mmol/mL and μmol/μL are both the same as mol/L or M. Other useful realizations include mM = μmol/mL.

ASSAY

An assay is the act of measuring how fast a given (or unknown) amount of enzyme will convert substrate to product—the act of measuring a velocity.

How fast a given amount of substrate is converted to product depends on how much enzyme is present. By measuring how much product is formed in a given time, the amount of enzyme present can be determined. An assay requires that you have some way to determine the concentration of product or substrate at a given time after starting the reaction. If the product and substrate have different UV or visible spectra, fluorescence spectra, and so forth, the progress of the reaction can be followed by measuring the change in the spectrum with time. If there are no convenient spectral changes, physical separation of substrate and product may be necessary. For example, with a radioactive substrate, the appearance of radioactivity in the product can be used to follow product formation.

VELOCITY

Velocity—*rate, v, activity, $d[P]/dt$, $-d[S]/dt$*—is how fast an enzyme converts substrate to product, the amount of substrate consumed or product formed per unit time. Units are micromoles per minute ($\mu mol/min$) = units.

There are a number of interchangeable words for velocity: the change in substrate or product concentration per time; rate; just plain v (for velocity, often written in italics to convince you it's special); activity; or the calculus equivalent, the first derivative of the product or substrate concentration with respect to time, $d[P]/dt$ or $-d[S]/dt$ (the minus means it's going away). Regardless of confusion, velocity (by any of its names) is just how fast you're going. Rather than miles per hour, enzyme velocity is measured in molar per minute (M/min) or more usually in micromolar per minute ($\mu M/min$).

Figure 7-1 shows what happens when enzyme is added to a solution of substrate. In the absence of enzyme, product appearance is slow, and there is only a small change in product concentration with time (low rate). After enzyme is added, the substrate is converted to product at a much

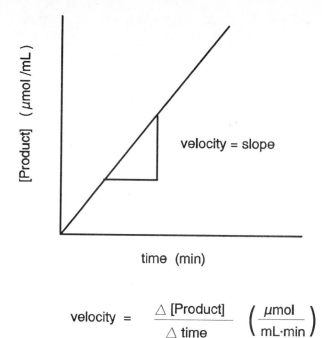

$$\text{velocity} = \frac{\triangle [\text{Product}]}{\triangle \text{ time}} \quad \left(\frac{\mu mol}{mL \cdot min} \right)$$

Figure 7-1
The **VELOCITY** of product formation (or substrate disappearance) is defined as
the change in product concentration per unit time. It is the slope of a plot of
product concentration against time. The velocity of product formation is the same
as the velocity of substrate disappearance (except that substrate goes away,
whereas product is formed).

faster rate. To measure velocity, you have to actually measure two things:
product (or substrate) concentration and time. You need a device to
measure concentration and a clock. Velocity is the *slope* of a plot of
product (or substrate) concentration (or amount) against time.
    Velocity can be expressed in a number of different units. The most
common is micromolar per minute (μM/min); however, because the velocity depends on the amount of enzyme used in the assay, the velocity is
often normalized for the amount of enzyme present by expressing the
activity in units of micromoles per minute per milligram of enzyme
[μmol/(min · mg)]. This is called a *specific activity*. You may be wondering (or not) where the volume went—after all, product concentration is
measured in molar units (M; mol/L). Well, it's really still there, but it
canceled out, and you don't see it because both the product and enzyme
are expressed in concentration units; both are per milliliter. Let's do a
numerical example.

Adding 0.1 μg of fumarase to a solution of 5 mM fumarate (substrate) in a final volume of 1.0 mL results in the formation of 0.024 μmol of malate (product) per minute. At the beginning of the reaction, the concentration of product is zero. If we wait 10 min, 10 × 0.024 μmol of malate will be made, or 0.24 μmol. In a volume of 1 mL, 0.24 μmol represents a concentration of 0.24 mM. So, over a period of 10 min, the concentration of malate went from 0 to 0.24 mM while the concentration of fumarate went from 5 to 4.76 mM (5 mM − 0.24 mM). The specific activity of the enzyme would be [0.024 μmol/(min · mL)]/(0.1 x 10^{-3} mg fumarase/mL) or 240 μmol/(min · mg).

To make matters worse, velocity is often reporting using the change in the amount of product per time (μmol/min). To actually determine the concentration, you need to know the volume. The key unit that always shows up somewhere with velocities and never cancels out is the *per time* part; the rest can usually be sorted out, depending on whether you're dealing with amounts or concentrations.

INITIAL VELOCITY

This is the measurement of the rate under conditions under which there is no significant change in the concentration of substrate.

As substrate is consumed, the substrate concentration falls and the reaction may get slower. As product is made, the reaction may slow down if the product is an inhibitor of the enzyme. Some enzymes are unstable and die as you're assaying them. All these things may cause the velocity to change with time. If the velocity is constant with time, the plot of product against time is a straight line; however, if velocity changes with time (the slope changes with time), this plot is curved (Fig. 7-2).

Usually enzyme activities are measured under conditions under which only a tiny bit of the substrate is converted to product (like 1 to 5 percent).[3] This means that the actual concentration of substrate will be very close to what you started with. It's obviously changed (or you couldn't have measured it), but it's not changed all that much. When you stick with the initial part of the velocity measurement, the velocity is less

---

[3] This is not a completely true statement. As you may see later on, the velocity of an enzyme-catalyzed reaction depends on the concentration of substrate only when the substrate concentration is near the K_m. If we start out with a concentration of substrate that is 1000 times the K_m, most of the substrate will have to be used up before the velocity falls because of a decrease in substrate concentration. If the product of the reaction does not inhibit and the enzyme is stable, the velocity will remain constant for much more than 1 to 5 percent of the reaction. It's only when we're near the K_m that substrate depletion during the assay is a problem.

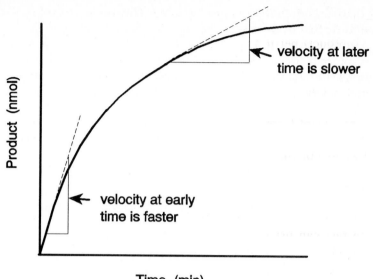

Figure 7-2 The Velocity May Change with Time
The velocity is not necessarily the same at all times after you start the reaction. The depletion of substrate, inhibition by the product, or instability of the enzyme can cause the velocity to change with time. The initial velocity is measured early, before the velocity changes. Initial velocity measurements also let you assume that the amount of substrate has not changed and is equal to the amount of substrate that was added.

likely to change due to substrate depletion, or product inhibition, and you're more likely to know what the actual substrate concentration is. Initial velocity measurements help avoid the dreaded curvature.

MECHANISM

$$E + S \rightleftharpoons ES \rightleftharpoons E + P$$

A mechanism tells you what happens to whom, in what order, and where. The mechanism in the box is the simplest one possible for a one-substrate enzyme. First, the enzyme must find the substrate in solution and bind to it, forming the ES complex in which the substrate is bound at the active site. The ES complex then converts the substrate to product and releases it. Real enzymes are obviously more complicated and almost never proceed through just one enzyme-substrate–enzyme-product complex—there are often many steps involved. However, even though there may be lots of steps involved, the slowest one will really determine the rate of the overall reaction. For this reason, the Michaelis-Menten mechanism and equation (see below) describe the behavior of a large number of enzymes. Many very complex mechanisms often follow the simple Michaelis-Menten equation when only one substrate is varied at a time and the others are held constant.

LITTLE k's

First-order: $A \longrightarrow B$ $v = k[A]$ k in s^{-1}
Second-order: $A + B \longrightarrow C$ $v = k[A][B]$ k in $M^{-1}s^{-1}$

The K_m and V_{max} of the Michaelis-Menten equation are actually made up of sums and products of little k's. You only have to look in most biochemistry texts to see a description of the derivation of the Michaelis-Menten equation in terms of little k's. The little k's are like quarks and leptons—you've heard the names, but you're not quite sure what they are and even less sure about how they work. There's a section later (actually last) in the book if you haven't heard or can't remember about rate constants.

The little k's are rate constants (numbers) that tell you how fast the individual steps are. You will see two kinds of little k's, rate constants—first-order and second-order (Fig. 7-3).

First-order rate constants are used to describe reactions of the type $A \rightarrow B$. In the simple mechanism for enzyme catalysis, the reactions

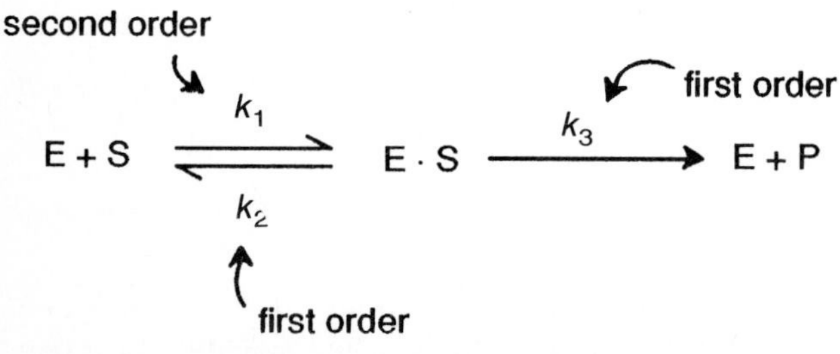

second order

Figure 7-3   The Simplest Enzyme Mechanism
A mechanism provides a description of individual chemical steps that make up the overall reaction. How fast each reaction occurs is governed by the rate constant for the reaction. The observable kinetic constants K_m and V_{max} are related to the individual rate constants for the individual steps by a bunch of algebra.

leading away from ES in both directions are of this type. The velocity of ES disappearance by any single pathway (like the ones labeled k_2 and k_3) depends on the fraction of ES molecules that have sufficient energy to get across the specific activation barrier (hump) and decompose along a specific route. ES gets this energy from collision with solvent and from thermal motions in ES itself. The velocity of a first-order reaction depends linearly on the amount of ES left at any time. Since velocity has units of molar per minute (M/min) and ES has units of molar (M), the little k (first-order rate constant) must have units of reciprocal minutes (1/min, or min^{-1}). Since only one molecule of ES is involved in the reaction, this case is called *first-order kinetics*. The velocity depends on the substrate concentration raised to the first power ($v = k[A]$).

The reaction of E with S is of a different type, called *second-order*. Second-order reactions are usually found in reactions of the type A + B $\rightarrow$ C. The velocity of a second-order reaction depends on how easy it is for E and S to find each other in the abyss of aqueous solution. Obviously, lower E or lower S concentrations make this harder. For second-order reactions, the velocity depends on the product of both of the reacting species ($v = k[S][E]$). Here k must have units of reciprocal molar minutes (M^{-1} min^{-1}) so that the units on the left and right sides balance. The second-order rate constant in the mechanism of Fig. 7-3 is k_1.

Another special case deserves comment—zero-order reactions. For a reaction that is zero-order with respect to a given substrate, the velocity does not depend on the concentration of substrate. We see zero-order behavior at V_{max}; the reaction is zero-order in the concentration of substrate. Note that even at V_{max}, the reaction is still first-order in the concentration of enzyme (the rate increases as the enzyme concentration increases).

MICHAELIS-MENTEN EQUATION

Hyperbolic kinetics, saturation kinetics

$$v = \frac{V_{max}[S]}{(K_m + [S])}$$

Velocity depends on substrate concentration when [S] is low but does not depend on substrate concentration when [S] is high.

Equations are useless by themselves—what's important is what the equation tells you about how the enzyme behaves. You should be able to escape biochemistry having learned only three equations. This is one of them. The Michaelis-Menten equation describes the way in which the velocity of an enzyme reaction depends on the substrate concentration. Memorize it and understand it because the same equation (with only the names of the symbols changed) describes receptor binding and oxygen binding to myoglobin. In the cell, the enzyme is exposed to changes in the level of the substrate. The way the enzyme behaves is to increase the velocity when the substrate concentration increases. The dependence of enzyme velocity on the concentration of substrate is the first line of metabolic regulation. In cells, enzymes are often exposed to concentrations of the substrate that are near to or lower than the K_m. Some enymes have evolved to have K_m's near the physiological substrate concentration so that changes in the substrate levels in the cell cause changes in the velocity of the reaction—the more substrate, the faster the reaction.

Mathematically, the Michaelis-Menten equation is the equation of a rectangular hyperbola. Sometimes you'll hear reference to *hyperbolic kinetics*; this means it follows the Michaelis-Menten equation. A number of other names also imply that a particular enzyme obeys the Michaelis-Menten equation: Michaelis-Menten behavior, saturation kinetics, hyperbolic kinetics.

The initial velocity is measured at a series of different substrate concentrations. In all cases the concentration of substrate used is much higher (by thousands of times, usually) than that of the enzyme. Each substrate concentration requires a separate measurement of the initial velocity. At low concentrations of substrate, increasing the substrate concentration increases the velocity of the reaction, but at high substrate concentrations, increasing the initial substrate concentration does not have much of an effect on the velocity (Fig. 7-4).

The Michaelis-Menten equation is one of those equations (like the Henderson-Hasselbalch) that required two people to come up with—and they still got it backward. The derivation of the equation is found in most texts and for the most part can be ignored. What you want to understand is how and why it works as it does.

$$v = \frac{d[P]}{dt} = -\frac{d[S]}{dt} = \frac{V_{max}[S]}{K_m + [S]}$$

When there is no substrate present ([S] = 0), there is no velocity—so far, so good. As the substrate concentration [S] is increased, the reaction goes faster as the enzyme finds it easier and easier to locate the substrate in solution. At low substrate concentrations ([S] $\ll K_m$), doubling the concentration of substrate causes the velocity to double.

As the initial substrate concentration becomes higher and higher, at some concentration, the substrate is so easy to find that all the enzyme

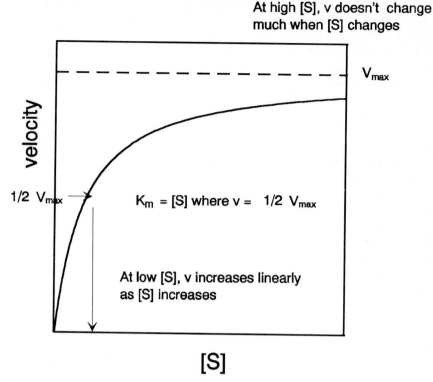

Figure 7-4
SUBSTRATE CONCENTRATION affects the velocity of an enzyme-catalyzed reaction. Almost all enzyme-catalyzed reactions show saturation behavior. At a high enough substrate concentration, the reaction just won't go any faster than V_{max}. The substrate concentration required to produce a velocity that is one-half of V_{max} is called the K_m.

active sites are occupied with bound substrate (or product). The enzyme is termed *saturated* at this point, and further increases in substrate concentration will not make the reaction go any faster. With $[S] \gg K_m$, the velocity approaches V_{max}.

The actual velocity of the reaction depends on how much of the total amount of enzyme is present in the enzyme–substrate (ES) complex. At low substrate concentrations, very little of the enzyme is present as the ES complex—most of it is free enzyme that does not have substrate bound. At very high substrate concentrations, virtually all the enzyme is in the ES complex. In fact, the amount of enzyme in the ES complex is just given by the ratio of v/V_{max}, that is, $ES/(ES+E) = v/V_{max} = [S]/(K_m + [S])$.

V_{max}

This is the velocity approached at a saturating concentration of substrate. V_{max} has the same units as v.

The V_{max} is a special point. At V_{max}, the velocity does not depend on the concentration of substrate. Most assays are performed at substrate concentrations that are near saturating (the word *near* is usually used because V_{max}, like Nirvana, is approached, not reached). For practical people, though, 99 percent of V_{max} is as good as V_{max}. The V_{max} and v have exactly the same units. The V_{max} conceals the dependence of the velocity on the concentration of enzyme. It's buried in there. If V_{max} is expressed in units of micromolar per minute ($\mu M/min$), then doubling the enzyme concentration doubles V_{max}; on the other hand, if V_{max} (and v) are given in units of micromoles per minute per milligram [$\mu mol/(min \cdot mg)$, i.e., specific activity], the normalized velocity and V_{max} won't depend on enzyme concentration.

---

k_{cat}

Turnover number—another way of expressing V_{max}
Micromoles of product made per minute per micromole of enzyme (V_{max}/E_t). The k_{cat} is the first-order rate constant for the conversion of the enzyme–substrate complex to product.

The turnover number, or k_{cat} (pronounced "kay kat"), is another way of expressing V_{max}. It's V_{max} divided by the total concentration of enzyme (V_{max}/E_t). The k_{cat} is a specific activity in which the amount of enzyme is expressed in micromoles rather than milligrams. The actual units of k_{cat} are micromoles of product per minute per micromole of enzyme. Frequently, the micromoles cancel (even though they're not exactly the same), to give you units of reciprocal minutes (min^{-1}). Notice that this has the same units as a first-order rate constant (see below, or see Chap. 22). The k_{cat} is the first-order rate constant for conversion of the enzyme–substrate complex to product. For a very simple mechanism, like the one shown above, k_{cat} would be equal to k_3. For more complex mechanisms k_{cat} is actually a collection of sums and products of rate constants for individual steps of the mechanism.

> # K_m
>
> This is the concentration of substrate required to produce a velocity that is one-half of V_{max}.

If $[S] = K_m$, the Michaelis-Menten equation says that the velocity will be one-half of V_{max}. (Try substituting $[S]$ for K_m in the Michaelis-Menten equation, and you too can see this directly.) It's really the relationship between K_m and $[S]$ that determines where you are along the hyperbola. Like most of the rest of biochemistry, K_m is backward. The larger the K_m, the weaker the interaction between the enzyme and the substrate. K_m is also a collection of rate constants. It may not be equal to the true dissociation constant of the ES complex (i.e., the equilibrium constant for $ES \rightleftharpoons E + S$).

> # SPECIAL POINTS
>
> $[S] \ll K_m$ $v = V_{max}[S]/K_m$
> $[S] \gg K_m$ $v = V_{max}$
> $[S] = K_m$ $v = V_{max}/2$

The K_m is a landmark to help you find your way around a rectangular hyperbola and your way around enzyme behavior. When $[S] \ll K_m$ (this means $[S] + K_m = K_m$), the Michaelis-Menten equation says that the velocity will be given by $v = (V_{max}/K_m)[S]$. The velocity depends linearly on $[S]$. Doubling $[S]$ doubles the rate.

At high substrate concentrations relative to K_m ($[S] \gg K_m$), The Michaelis-Menten equation reduces to $v = V_{max}$, substrate concentration disappears, and the dependence of velocity on substrate concentration approaches a horizontal line. When the reaction velocity is independent of the concentration of the substrate, as it is at V_{max}, it's given the name *zero-order kinetics*.

> # k_{cat}/K_m
>
> The specificity constant, k_{cat}/k_m, is the second-order rate constant for the reaction of E and S to produce product. It has units of $M^{-1}min^{-1}$.

The term V_{max}/K_m describes the reaction of an enzyme and substrate at low substrate concentration. At low substrate concentration, the velocity of an enzyme-catalyzed reaction is proportional to the substrate concentration and the enzyme concentration. The proportionality constant is k_{cat}/K_m and $v = (k_{cat}/K_m)[S] [E]_T$. If you're real astute, you'll have noticed that this is just a second-order rate equation and that the second-order rate constant is k_{cat}/K_m.

The term k_{cat}/K_m lets you rank enzymes according to how good they are with different substrates. It contains information about how fast the reaction of a given substrate would be when it's bound to the enzyme (k_{cat}) and how much of the substrate is required to reach half of V_{max}. Given two substrates, which will the enzyme choose? The quantity k_{cat}/K_m tells you which one the enzyme likes most—which one will react faster.

The term k_{cat}/K_m is also the second-order rate constant for the reaction of the free enzyme (E) with the substrate (S) to give product. The k_{cat}/K_m is a collection of rate constants, even for the simple reaction mechanism shown above. Formally, k_{cat}/K_m is given by the pile of rate constants $k_1 k_3/(k_2 + k_3)$. If $k_3 \gg k_2$, this reduces to k_1, the rate of encounter between E and S. Otherwise, k_{cat}/K_m is a complex collection of rate constants, but it is still the second-order rate constant that is observed for the reaction at low substrate concentration.

RATE ACCELERATIONS[4]

Compare apples to apples and first-order to first-order.

To impress you, enzymologists often tell you how much faster their enzyme is than the uncatalyzed reaction. These comparisons are tricky. Here's the problem: Suppose we know that the reaction $S \rightarrow P$ has a first-order rate constant of 1×10^{-3} min^{-1} (a half-life of 693 min). When an enzyme catalyzes the transformation of S to P, we have more than one reaction:

$$E + S \underset{k_2}{\overset{k_1}{\rightleftharpoons}} ES \xrightarrow{k_3} E + P$$

Which of these three reactions do we pick to compare with the noncatalyzed reaction? We can't pick k_1, because that's a second-order

---

[4] If you're reading this section because you want to understand how rate accelerations are actually determined, proceed; however, this information will be pretty low on the trivia sorter list.

reaction. You can't directly compare first- and second-order reactions—
the units are different. The comparison to make in this case is with k_3.
You're actually comparing the first-order rate constant for the reaction
when the substrate is free in solution to the first-order rate constant for
the reaction when the substrate is bound to the enzyme. So, for a first-
order reaction, we compare the uncatalyzed rate constant to the k_{cat} for
the enzyme-catalyzed reaction. If k_{cat} were 10^3 min^{-1} (half-life of
0.000693 min, or 41 ms), the rate acceleration would be 10^6-fold.

What about reactions of the type A + B → C? This is a second-order
reaction, and the second-order rate constant has units of M^{-1}min^{-1}. The
enzyme-catalyzed reaction is even more complicated than the very simple
one shown above. We obviously want to use a second-order rate constant
for the comparison, but which one? There are several options, and all
types of comparisons are often made (or avoided). For enzyme-catalyzed
reactions with two substrates, there are two K_m values, one for each
substrate. That means that there are two k_{cat}/K_m values, one for each
substrate. The k_{cat}/K_A[5] in this case describes the second-order rate constant for the reaction of substrate A with whatever form of the enzyme
exists at a saturating level of B. Cryptic enough? The form of the enzyme
that is present at a saturating level of B depends on whether or not B can
bind to the enzyme in the absence of A.[6] If B can bind to E in the absence
of A, then k_{cat}/K_A will describe the second-order reaction of A with the
EB complex. This would be a reasonably valid comparison to show the
effect of the enzyme on the reaction. But if B can't bind to the enzyme in
the absence of A, k_{cat}/K_A will describe the second-order reaction of A
with the enzyme (not the EB complex). This might not be quite so good a
comparison.

STEADY-STATE APPROXIMATION

This is an assumption used to derive the Michaelis-Menten equation
in which the velocity of ES formation is assumed to be equal to the
velocity of ES breakdown.

As with most assumptions and approximations, those professors who
do not ignore this entirely will undoubtedly think that you should at least
know that it's an assumption. What the steady-state assumption actually

---

[5] K_A is the K_m for substrate A determined at a saturating level of substrate B.

[6] Sometimes the binding site for one substrate does not exist until the other substrate binds
to the enzyme. This creates a specific binding order in which A must bind before B can bind
(or vice versa).

does is to allow enzyme kineticists (a hale, hardy, wise, but somewhat strange breed) to avoid calculus and differential equations. Think about what happens when E and S are mixed for the first time. E begins to react with S. Although the S concentration is large and constant, the E concentration begins to drop as it is converted to ES ($v = k_1[E][S]$). At the same time, the velocity of ES breakdown to E + S and to E + P starts to rise as the ES concentration increases. Since ES is destroyed by two different reactions, the velocity of ES breakdown is the sum of the velocities of the two pathways ($v = k_2[ES] + k_3[ES]$). At some point, the concentrations of E and ES will be just right, and the velocity at which ES is created will be exactly matched by the velocity at which ES is converted to other things. As long as the velocity of ES formation remains the same as the velocity of ES destruction, the concentration of ES will have to stay constant with time. At this point the system has reached steady state. At steady state, the concentration of ES (and other enzyme species) won't change with time (until [S] decreases, but we won't let that happen because we're measuring initial velocity kinetics). If the concentration of the ES complex doesn't change with time, it really means that $d[ES]/dt = 0$. This has consequences.

At any time, the velocity of product formation is

$$v = k_3[ES]$$

The k_3 is just a constant. What we don't know is [ES], mainly because we don't know how much of our enzyme is present as [E]. But we do know how much total enzyme we have around—it's how much we added.

$$[E]_{total} = [E] + [ES]$$

But we still don't know [ES]. We need another equation that has [E] and [ES] in it. Here's where the steady state approximation comes in handy. At steady state, the change in the concentration of [ES] is zero, and the velocity of [ES] formation equals the velocity of [ES] breakdown.

$$
\begin{aligned}
v_{formation} &= k_1[E] \\
v_{breakdown} &= k_2[ES] + k_3[ES] \\
k_1[E] &= (k_2 + k_3)[ES]
\end{aligned}
$$

The rest is simple algebra.[7] Solve the equation above for [E] and stick the

---

[7] At least, it's simpler than calculus.

result in the equation for $[E]_{total}$. The E and ES should disappear at this point, and you should be left with something that has only k's and $[E]_{total}$ in it. Call $k_3[E]_t$ by the name V_{max} and call $(k_2 + k_3)/k_1$ by the name K_m, and you've just derived the Michaelis-Menten equation.

TRANSFORMATIONS AND GRAPHS

$1/v$ vs. $1/[S]$	Lineweaver-Burk
v vs. $v/[S]$	Eadie-Hofstee
$[S]/v$ vs. $[S]$	Hanes-Wolf

Confirming that curved lines are the nemesis of the biochemist, at least three or more different transformations of the Michaelis-Menten equation have been invented (actually four)—each one of which took two people to accomplish (Fig. 7-5). The purpose of these plots is to allow you to determine the values of K_m and V_{max} with nothing but a ruler and a piece of paper and to allow professors to take a straightforward question about the Michaelis-Menten expression and turn it upside down (and/or backward). You might think that turning a backward quantity like K_m upside down would make everything simpler—somehow it doesn't work that way.

Don't bother memorizing these equations—they're all straight lines with an x and y intercept and a slope. The useful information (V_{max} or K_m) will either be on the x and y intercepts or encoded in the slope (y intercept/x intercept). Remember that v has the same units as V_{max}, and K_m has the same units as $[S]$. Look at the label on the axes and it will tell you what the intercept on this axis should give in terms of units. Then match these units with the units of V_{max} and $[S]$.

The y intercept of a Lineweaver-Burk plot ($1/v$ is the y axis) is $1/V_{max}$ (same units as $1/v$). The x intercept has the same units as $1/[S]$ so that $1/K_m$ (actually $-1/K_m$) is the x intercept. For the Eadie-Hofstee plot, the x axis is $v/[S]$ so that the x intercept is V_{max}/K_m (units of the x axis are $v/[S]$ units). Get the idea?

The useful thing about the Lineweaver-Burk transform (or double reciprocal) is that the y intercept is related to the first-order rate constant for decomposition of the ES complex to E + P (k_{cat} or V_{max}) and is equal to the rate observed with all of the enzyme in the ES complex. The slope, on the other hand, is equal to the velocity when the predominant form of the enzyme is the *free* enzyme, E (*free* meaning unencumbered rather than cheap).

LINEWEAVER-BURK

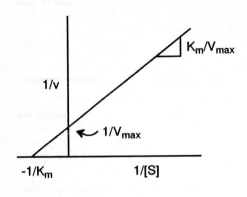

EADIE-HOFSTEE

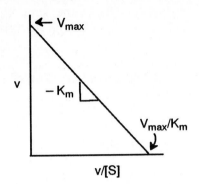

HANES-WOLF

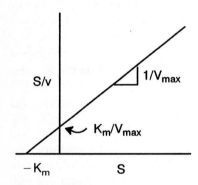

Figure 7-5
TO AVOID CURVED PLOTS there are several ways to rearrange the Michaelis-Menten equation so that data can be plotted to give a straight line. The slope and intercepts give you values for K_m and V_{max} or give you values from which K_m and V_{max} can be calculated.

INHIBITION

Competitive: Slope effect
Uncompetitive: Intercept effect
Noncompetitive: Slope and intercept effect

Inhibitors are molecules that often resemble the substrate(s) or product(s) and bind reversibly to the active site (this means if you remove the inhibitor, the inhibition goes away). Binding of the inhibitor to the active site prevents the enzyme from turning over.[8] Many drugs are reversible enzyme inhibitors that have their physiological effect by decreasing the activity of a specific enzyme. If an inhibitor has an effect on the velocity, it will be to decrease the velocity. The concentration of inhibitor needed to inhibit the enzyme depends on how tightly the inhibitor binds to the enzyme. The *inhibition constant* (K_i) is used to describe how tightly an inhibitor binds to an enzyme. It refers to the equilibrium dissociation constant of the enzyme–inhibitor complex (equilibrium constant for $EI \rightleftharpoons E + I$). The bigger the K_i, the weaker the binding. (At least it's consistently backward.)

There are three types of reversible inhibition: competitive, uncompetitive, and noncompetitive. Most texts acknowledge only two kinds of inhibition—competitive and noncompetitive (or mixed). This approach makes it difficult to explain inhibition on an intuitive level, so we'll use all three types of inhibition and explain what the other nomenclature means in the last paragraph of this section.

Inhibition experiments are performed by varying the concentration of substrate around the K_m just like an experiment to determine the K_m and V_{max}, except that the experiment is repeated at several different concentrations of an inhibitor. On a Lineweaver-Burk transformation ($1/v$ vs. $1/[S]$), each different inhibitor concentration will be represented as a different straight line (Fig. 7-6). The pattern that the lines make tells you the kind of inhibition. There are three possibilities: (1) The inhibitor can affect only the slopes of the plot (competitive), (2) the inhibitor can affect only the y intercepts of the plot (uncompetitive), or (3) the inhibitor can affect both the slopes and the intercepts (noncompetitive). Plots are plots, and what's really important is not the pattern on a piece of paper but what the pattern tells us about the behavior of the enzyme.

The Lineweaver-Burk plot is very useful for our purposes since the y axis intercept of this plot ($1/V_{max}$) shows the effect of the inhibitor at a very high concentration of the substrate. On the other hand, the slope of this plot (K_m/V_{max}) shows the effect of the inhibitor at a very low concentration of substrate.

The hallmarks of competitive inhibition are that V_{max} is not affected by adding the inhibitor and the plots intersect on the y axis. A high concentration of substrate prevents the inhibitor from exerting its effect.

---

[8]An enzyme's turning over is not like a trick your dog can do. *Enzyme turnover* refers to the cyclic process by which the enzyme turns the substrate over into product.

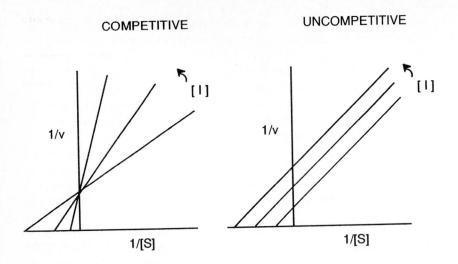

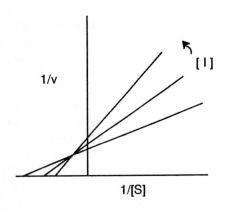

Figure 7-6   Enzyme Inhibition
An inhibitor can have different effects on the velocity when the substrate concentration is varied. If the inhibitor and substrate compete for the same form of the enzyme, the inhibition is **COMPETITIVE.** If not, the inhibition is either **NONCOMPETITIVE** or **UNCOMPETITIVE** depending on whether or not the inhibitor can affect the velocity at low substrate concentrations.

Competitive inhibitors bind only to the free enzyme and to the same site as the substrate. Competitive inhibitors are molecules that usually look like the substrate but can't undergo the reaction. At an infinite concentration of the substrate ($1/[S] = 0$), the competitive inhibitor cannot bind to the enzyme since the substrate concentration is high enough that there is virtually no free enzyme present.

COMPETITIVE

$$E + S \rightleftharpoons ES \longrightarrow E + P$$
$$\Updownarrow \pm I$$
$$EI$$

Since competitive inhibitors have no effect on the velocity at saturating (infinite) concentrations of the substrate, the intercepts of the double-reciprocal plots ($1/V_{max}$) at all the different inhibitor concentrations are the same. The lines at different inhibitor concentrations must all intersect on the y axis at the same $1/V_{max}$.

At low concentrations of substrate ($[S] \ll K_m$), the enzyme is predominantly in the E form. The competitive inhibitor can combine with E, so the presence of the inhibitor decreases the velocity when the substrate concentration is low. At low substrate concentration ($[S] \ll K_m$), the velocity is just $V_{max}[S]/K_m$. Since the inhibitor decreases the velocity and the velocity at low substrate concentration is proportional to V_{max}/K_m, the presence of the inhibitor affects the slopes of the Lineweaver-Burk plots; the slope is just the reciprocal of V_{max}/K_m. Increasing the inhibitor concentration causes K_m/V_{max} to increase. The characteristic pattern of competitive inhibition can then be rationalized if you simply remember that a competitive inhibitor combines only with E.

If the inhibitor combines only with ES (and not E), the inhibitor exerts its effect only at high concentrations of substrate at which there is lots of ES around. This means that the substrate you're varying (S) doesn't prevent the binding of the inhibitor and that the substrate and inhibitor bind to two different forms of the enzyme (E and ES, respectively).

UNCOMPETITIVE

$$E + S \rightleftharpoons ES \longrightarrow E + P$$
$$\Updownarrow \pm I$$
$$ESI$$

At very low substrate concentration ([S] approaches zero), the enzyme is mostly present as E. Since an uncompetitive inhibitor does not combine with E, the inhibitor has no effect on the velocity and no effect on V_{max}/K_m (the slope of the double-reciprocal plot). In this case, termed *uncompetitive*, the slopes of the double-reciprocal plots are independent of inhibitor concentration and only the intercepts are affected. A series of parallel lines result when different inhibitor concentrations are used. This type of inhibition is often observed for enzymes that catalyze the reaction between two substrates. Often an inhibitor that is competitive against one of the substrates is found to give uncompetitive inhibition when the other substrate is varied. The inhibitor does combine at the active site but does not prevent the binding of one of the substrates (and vice versa).

NONCOMPETITIVE

$$E + S \rightleftharpoons ES \longrightarrow E + P$$

$$\Updownarrow \pm I \qquad \Updownarrow \pm I$$

$$EI \qquad\qquad ESI$$

Noncompetitive inhibition results when the inhibitor binds to both E and ES. Here, both the slopes (K_m/V_{max}) and intercepts ($1/V_{max}$) exhibit an effect of the inhibitor. The lines of different inhibitor concentration intersect (their slopes are different), but they do not intersect on the y axis (their intercepts are different).

In many texts, the existence of uncompetitive inhibitors is ignored, and noncompetitive inhibitors are called *mixed*. The different types of inhibitors are distinguished by their effects of K_m and V_{max}. It is difficult to rationalize the behavior of inhibitors by discussing their effects on K_m; however, it is well suited for memorization. In this terminology, noncompetitive inhibition is observed when the inhibitor changes V_{max} but doesn't have an effect on K_m. The lines intersect at a common K_m on the x axis. In true noncompetitive inhibition, the inhibitor does not effect K_m only if the dissociation constants of the inhibitor from the EI and ESI are both the same. Otherwise, the point of intersection is above or below the x axis depending on which of the two inhibition constants is bigger. The term *mixed inhibition* is used to refer to noncompetitive inhibition in which both K_m and V_{max} are affected by the inhibitor. This is all terribly and needlessly confusing. If you're not going to be an enzymologist, the best advice is to just give up.

ALLOSTERISM AND COOPERATIVITY

Cooperativity: Observed when the reaction of one substrate molecule with a protein has an effect on the reaction of a second molecule of the substrate with another active site of the protein.

Positive: The binding of the first substrate makes the reaction of the next substrate *easier*.

Negative: The binding of the first substrate makes the reaction of the next substrate *harder*.

Allosterism: The binding of an effector molecule to a separate site on the enzyme affects the K_m or V_{max} of the enzyme.

Allosterism and cooperativity are lumped together because of the commonality of the structural changes required by both. The essence of both effects is that binding (and catalytic) events at one active site can influence binding (and catalytic) events at another active site in a multimeric protein. Cooperativity requires a protein with multiple active sites. They are usually located on multiple subunits of a multimeric protein. Cooperative enzymes are usually dimers, trimers, tetramers, and so forth. This implies that the binding of the effector molecule (the one causing the effect) changes the structure of the protein in a way that tells the other subunits that it has been bound. Frequently, the regulatory regions and active sites of allosteric proteins are found at the interface regions between the subunits.

The separation between allosteric effectors and cooperativity lies in the molecule doing the affecting. If the effector molecule acts at another site and the effector is not the substrate, the effect is deemed *allosteric* and *heterotropic*. If the effector molecule is the substrate itself, the effect is called *cooperative* and/or *homotropic*.

Positive cooperativity means that the reaction of substrate with one active site makes it easier for another substrate to react at another active site. Negative cooperativity means that the reaction of a substrate with one active site makes it harder for substrate to react at the other active site(s).

Cooperative enzymes show sigmoid or sigmoidal kinetics because the dependence of the initial velocity on the concentration of the substrate is not Michaelis-Menten-like but gives a *sigmoid* curve (Fig. 7-7).

Since the effects of substrate concentration on the velocity of a cooperative enzyme are not described by a hyperbola (Michaelis-Menten), it's not really appropriate to speak of K_m's. The term reserved for

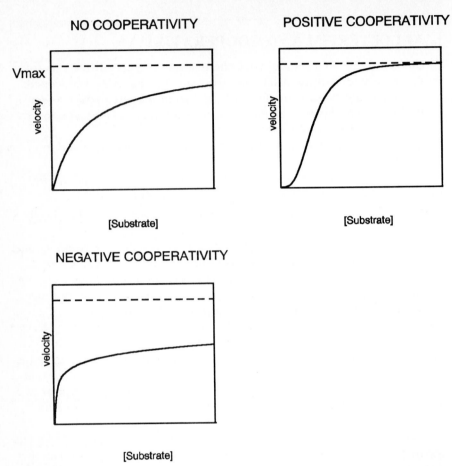

Figure 7-7
COOPERATIVE ENZYMES do not show a hyperbolic dependence of the velocity on substrate concentration. If the binding of one substrate increases the affinity of an oligomeric enzyme for binding of the next substrate, the enzyme shows positive cooperativity. If the first substrate makes it harder to bind the second substrate, the enzyme is negatively cooperative.

the special concentration of substrate that produces a velocity that is one-half of V_{max} is $S_{0.5}$. Enzymes that are positively cooperative are very sensitive to changes of substrate near the $S_{0.5}$. This makes the enzyme behave more like an on–off switch and is useful metabolically to provide a large change in velocity in response to a small change in substrate concentration. Negative cooperativity causes the velocity to be rather insensitive to changes in substrate concentration near the $S_{0.5}$.

THE MONOD-WYMAN-CHANGEAUX MODEL

The MWC model describes cooperativity.

Each subunit exists in a conformational state that has either a low affinity (T state) or a high affinity (R state) for substrate.

In any one enzyme molecule, all the subunits are in the same conformational state.

In the absence of substrate, the T state is favored. In the presence of substrate, the R state is favored.

The model most often invoked to rationalize cooperative behavior is the MWC (Monod-Wyman-Changeaux), or concerted, model. This model is 1.5 times more complicated that the Michaelis-Menten model and took three people to develop instead of two. Most texts describe it in detail. In the absence of substrate, the enzyme has a low affinity for substrate. The MWC folks say that the enzyme is in a T (for *tense* or *taut*) state in the absence of substrate. Coexisting with this low-affinity T state is another conformation of the enzyme, the R (for *relaxed*) state, that has a higher affinity for substrate. The T and R states coexist in the absence of substrate, but there's much more of the T state than the R. This has always seemed backward, since one would expect the enzyme to be more tense in the presence of substrates when some work is actually required. In keeping with the tradition of biochemistry, the MWC folks obviously wanted this to be backward too (Fig. 7-8).

The MWC model says that in the R state, all the active sites are the same and all have higher substrate affinity than in the T state. If one site is in the R state, all are. In any one protein molecule at any one time, all subunits are supposed to have identical affinities for substrate. Because the transition between the R and the T states happens at the same time to all subunits, the MWC model has been called the *concerted model* for allosterism and cooperativity. The MWC model invokes this symmetry principle because the modelers saw no compelling reason to think that one of the chemically identical subunits of a protein would have a conformation that was different from the others. Alternative models exist that suggest that each subunit can have a different conformation and different affinities for substrate. Experimentally, examples are known that follow each model.

The arithmetic of the MWC model is not worth going into, but the sigmoidal behavior arises from the fact that the enzyme is capable of interacting with multiple ligands with reactions of the type ($E + 4S \rightleftharpoons$

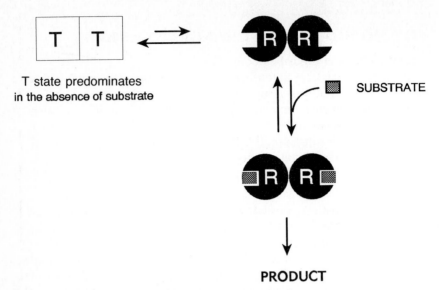

T state predominates
in the absence of substrate

SUBSTRATE

PRODUCT

Figure 7-8   The MWC MODEL for Positive Cooperativity
In the absence of substrate, there is more of the T state than the R state. Substrate
binds more tightly to the R state. Within one enzyme molecule, the subunits are all
T or all R.

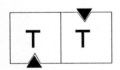

**Allosteric inhibitors
bind specifically
to the T state and
make it harder for
substrate to switch
enzyme into the
R state**

**Allosteric activators
bind specifically to the
R state and pull more
of the enzyme into the
more active R state.**

Figure 7-9
ALLOSTERIC EFFECTORS bind specifically to either the T or the R states.
Heterotropic (nonsubstrate) activators bind to and stabilize the R state, while
heterotropic inhibitors bind preferentially to the T state.

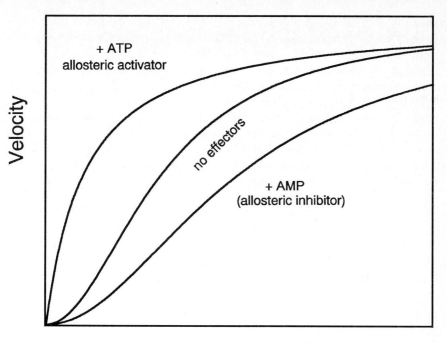

[Fructose-6-Phosphate]

Figure 7-10
PHOSPHOFRUCTOKINASE shows positive cooperativity with fructose-6-phosphate as the substrate. ATP, an allosteric inhibitor, binds to the T state and decreases the velocity. AMP, a signal for low energy, binds to the R state and increases the velocity of the reaction.

ES_4). This type of binding introduces exponents into the concentration of substrate terms so that you can draw a curve like the one shown in Fig. 7-8.

A substrate or effector that binds preferentially to the R state increases the concentration of the R state at equilibrium. This can only happen if, in the absence of substrate or effector, the enzyme is predominantly in the T state. If the enzyme were predominantly in the R state to begin with, it would already have increased affinity for the substrate and there would be no allosteric or cooperative effects. Consequently, the MWC model cannot account for negative cooperativity (but this is rare anyway).

Substrates can affect the conformation of the other active sites. So can other molecules. Effector molecules other than the substrate can bind to specific effector sites (different from the substrate-binding site) and

shift the original T–R equilibrium (Fig. 7-9). An effector that binds preferentially to the T state decreases the already low concentration of the R
state and makes it even more difficult for the substrate to bind. These
effectors decrease the velocity of the overall reaction and are referred to
as *allosteric inhibitors*. An example is the effect of ATP or citrate on the
activity of phosphofructokinase. Effectors that bind specifically to the R
state shift the T–R equilibrium toward the more active (higher-affinity) R
state. Now all the sites are of the high-affinity R type, even before the first
substrate binds. Effectors that bind to the R state increase the activity
(decrease the $S_{0.5}$) and are known as *allosteric activators*. An example is
the effect of AMP or ATP on the velocity of the phosphofructokinase
reaction (Fig. 7-10). Notice that in the presence of an allosteric activator,
the *v* versus [S] plot looks more hyperbolic and less sigmoidal—consistent with shifting the enzyme to all R-type active sites.

The effects of substrates and effectors have been discussed with
regard to how they could change the affinity of the enzyme for substrate
($S_{0.5}$). It would have been just as appropriate to discuss changes in V_{max}
in a cooperative, or allosteric, fashion.

· C H A P T E R · 8 ·

GLYCOLYSIS AND GLUCONEOGENESIS

·

Glycolysis Function

Glycolysis Location

Glycolysis Connections

Glycolysis Regulation

Glycolysis ATP Yields

Glycolysis Equations

Effect of Arsenate

Lactate or Pyruvate

Gluconeogenesis Function

Gluconeogenesis Location

Gluconeogenesis Connections

Gluconeogenesis Regulation

Gluconeogenesis ATP Costs

Gluconeogenesis Equations

· · · · · · · · · · · ·

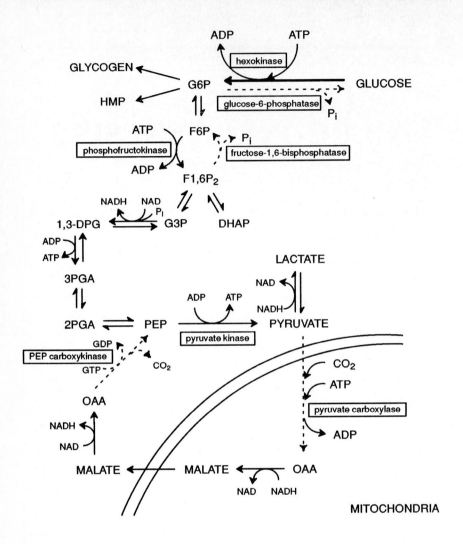

Figure 8-1
GLYCOLYSIS (solid lines) and **GLUCONEOGENESIS** (dotted lines) share some
common enzymes, but they diverge around the control steps. Major control
enzymes are boxed. Signals that turn glycolysis on turn gluconeogenesis off, and
vice versa.

GLYCOLYSIS FUNCTION

Aerobic: To convert *glucose* to pyruvate and ATP. Pyruvate can be burned for energy (TCA) or converted to fat (fatty acid synthesis).
Anaerobic: ATP production. Recycle NADH by making lactate.

GLYCOLYSIS LOCATION

Everywhere

GLYCOLYSIS CONNECTIONS

Glucose *in*, pyruvate or lactate *out*
Glucose 6-phosphate to glycogen (reversible)
Glucose 6-phosphate to pentose phosphates (*not* reversible)
Pyruvate to TCA via acetyl-CoA (*not* reversible)
Pyruvate to *fat* via acetyl-CoA (*not* reversible)

GLYCOLYSIS REGULATION

Primary signals: *Insulin* turns *on*.
                       Glucagon turns *off*.
                       Epinephrine turns *on* in *muscle, off* in *liver*.
                       Phosphorylation turns *off* in *liver, on* in *muscle*.
Secondary signals: Glucose signals activate
                        (fructose 2,6-bisphosphate activates
                        phosphofructokinase).
                        Low-glucose signals inhibit.
                        High-energy signals inhibit.
                        Low-energy signals activate.

GLYCOLYSIS ATP YIELDS

Aerobic: 1 glucose $\longrightarrow$ 2ATP + 2NADH + 2 pyruvate

Anaerobic: 1 glucose $\longrightarrow$ 2ATP + 2 lactate

Complete: 1 glucose $\longrightarrow$ $6CO_2$ + 38ATP

(using malate/aspartate shuttle to oxidize NADH)

From glycogen: $(Glycogen)_n \longrightarrow$ 3ATP + 2NADH + 2 pyruvate

GLYCOLYSIS EQUATIONS

Aerobic: $Glucose + 2ADP + 2P_i + 2NAD^+ \longrightarrow$
$2\ pyruvate + 2ATP + 2NADH + 2H^+$

Anaerobic: $Glucose + 2ADP + 2P_i \longrightarrow 2\ lactate + 2ATP$

EFFECT OF ARSENATE

This is a common exam question.

Arsenate ($HAsO_4^{-2}$), an analog of phosphate, has an interesting effect on glycolysis. This makes a great exam question. Arsenate is a substrate for the enzyme glyceraldehyde-3-phosphate dehydrogenase. The enzyme, which normally uses phosphate and makes 1,3-diphosphoglycerate, is fooled by the arsenate and makes the arsenate ester instead. With the phosphate ester, the next enzyme in glycolysis makes an ATP from the 1,3-diphosphoglycerate. The arsenate analog of 1,3-diphosphoglycerate is chemically much more unstable than the phosphate ester and hydrolyzes to 3-phosphoglycerate before an ATP can be made. The product, however, is the same, 3-phosphoglycerate, so glycolysis can continue as normal. But what has happened is that this step no longer makes an ATP for each three-carbon fragment. You lose 2 ATPs per glucose—all the net ATP production of glycolysis. The bottom line is that glycolysis continues (in fact it's usually accelerated by the lack of ATP), but no ATP can be made. It is very analogous to the uncoupling of oxidative phosphorylation by dinitrophenol.

LACTATE OR PYRUVATE

With oxygen present, pyruvate is oxidized by the tricarboxylic acid
    cycle (TCA).
Without oxygen, pyruvate is reduced to lactate.
In muscle, lactate is usual product.

The product of glycolysis is pyruvate. The pyruvate made by glycolysis can either enter the TCA cycle through pyruvate dehydrogenase or be reduced to lactate. To keep running, glycolysis requires NAD^+ in the glyceraldehyde-3-phosphate dehydrogenase reaction. No NAD^+, no glycolysis. NAD^+ is produced by oxidation of NADH via oxidative phosphorylation, a process that requires oxygen. Under anaerobic conditions, the cycle simply shuts down and the pyruvate must be converted to lactate to keep glycolysis going. In muscle, glycolysis is normally faster than the TCA cycle capacity, and lactate is the usual product of glycolysis even in resting muscle. The lactate/pyruvate ratio is about 10 in resting muscle, but in working muscle this ratio may hit 200.

GLUCONEOGENESIS FUNCTION

Gluconeogenesis makes *glucose* from pyruvate to help maintain blood glucose levels.

GLUCONEOGENESIS LOCATION

Liver and kidney—*not muscle*

GLUCONEOGENESIS CONNECTIONS

Pyruvate in, glucose out
Lactate in, glucose out
Alanine in, glucose and urea out

   Gluconeogenesis in the liver can be fueled by molecules other than pyruvate or lactate. Alanine, a product of protein degradation, yields pyruvate by simple transamination, and this pyruvate can be converted to glucose by the liver and kidney. Other amino acids are metabolized to pyruvate or oxaloacetate, which can also enter the gluconeogenic pathway. In addition, glycerol from the breakdown of triglycerides in adipose tissue can be used by the liver and kidney to make glucose.[1]

GLUCONEOGENESIS REGULATION

Primary signals: Insulin turns *off*.
                         Glucagon turns *on*.
                         Acetyl-CoA turns *on*.
                         Phosphorylation turns *on* in liver.
Secondary signals: Glucose signals turn *off*.
                         (Fructose 2,6-bisphosphate inhibits
                         fructose 1,6-bisphosphatase.)
                         Low-glucose signals activate.
                         High-energy signals activate.
                         Low-energy signals inhibit.

   There are two unusual aspects to the regulation of gluconeogenesis. The first step in the reaction, the formation of oxaloacetate from pyruvate, requires the presence of acetyl-CoA. This is a check to make sure that the TCA cycle is adequately fueled. If there's not enough acetyl-CoA around, the pyruvate is needed for energy and gluconeogenesis won't happen. However, if there's sufficient acetyl-CoA, the pyruvate is shifted toward the synthesis of glucose.

GLUCONEOGENESIS ATP COSTS

2 lactate + 6ATP (equivalents) $\longrightarrow$ glucose

---

[1] The glycerol produced by the action of hormone-sensitive lipase in the adipose tissue cannot be utilized by adipose tissue itself. Adipose cells lack the enzyme glycerol kinase, which is necessary to convert glycerol to glycerol phosphate.

GLUCONEOGENESIS EQUATIONS

2 lactate + 4ATP + 2GTP $\longrightarrow$ glucose + 4ADP + 2GDP + 6P$_i$

2 pyruvate + 4ATP + 2GTP + 2NADH + 2H$^+$ $\longrightarrow$
$$4ADP + 2GDP + 6P_i + 2NAD^+$$

GLYCOGEN SYNTHESIS AND DEGRADATION

·

Function

Location

Connections

Regulation

ATP Yield

ATP Cost

Molecular Features

· · · · · · · · · · · ·

FUNCTION

To store *glucose* equivalents and retrieve them on demand

LOCATION

Major deposits in liver for maintaining blood glucose
Deposits in muscle for providing glucose for muscle energy requirements

LONG FORM

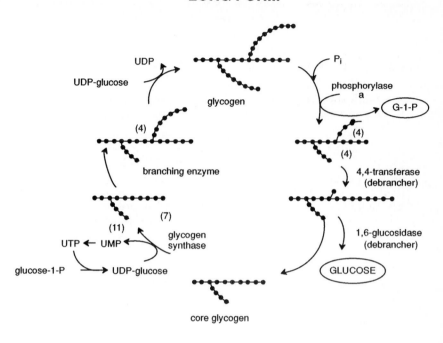

SHORT FORM

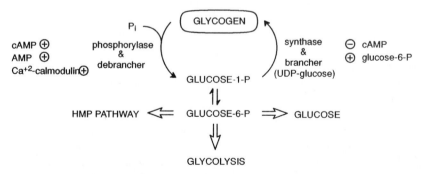

Figure 9-1 Glycogen Synthesis and Degradation
The short form shows the major control features. The long form indicates the
number of glucose residues required around the branch points to make the various
synthesis and degradation steps work correctly.

CONNECTIONS

Glycogen to and from *glucose 1-phosphate*
Glucose 1-phosphate to glucose 6-phosphate
Glucose 6-phosphate to *glucose (liver and kidney only)*
Glucose 6-phosphate from *glucose*
Glucose 6-phosphate to and from *glycolysis* and *gluconeogenesis*
Glucose 6-phosphate to pentose phosphates (*not* reversible)

REGULATION

Primary signals: *Insulin* turns *synthesis on, degradation off.*
Glucagon turns *synthesis off, degradation on.*
Epinephrine turns *synthesis off, degradation on.*
Phosphorylation turns *synthesis off, degradation on.*

Secondary signals: Glucose 6-phosphate activates *synthesis.*
Ca^{2+}-Calmodulin activates *degradation* by activating phosphorylase kinase.

The synthesis and degradation of glycogen provide control of the availability of glucose equivalents. Conditions that reflect low-glucose and/or low-energy levels turn on glycogen degradation and turn off glycogen synthesis. Regulation is principally through a cascade of phosphorylation that begins with increases in the concentration of cAMP brought about by the stimulation of adenylate cyclase by hormones for low-glucose (glucagon) and low-energy (epinephrine) levels. Glycogen phosphorylase, the enzyme that degrades glycogen to glucose 1-phosphate, is activated through phosphorylation catalyzed by phosphorylase kinase. The phosphorylase kinase is, in turn, activated by cAMP-dependent protein kinase. In the absence of cAMP signals, the activity of protein phosphatases keeps phosphorylase inactive and activates glycogen synthase. Glycogen synthesis is inactivated by phosphorylation of glycogen synthase, the enzyme responsible for making glycogen.

Regulation of glycogen synthesis and degradation is essentially the same in the liver and muscle, but there are a couple of wrinkles. Glycogen degradation is also activated in muscle in response to the rise in intracellular calcium levels that accompanies contraction. This is achieved by a stimulation of phosphorylase kinase that occurs when calmodulin (a regulatory protein associated with phosphorylase and some other proteins) binds calcium. In addition, glycogen synthesis can be activated by high levels of glucose 6-phosphate. Glycogen synthase, even when it's phosphorylated and inactive, can be stimulated by glucose 6-phosphate.

ATP YIELD

No ATP is required to remove glucose from glycogen stores.

Degradation:[1]

$(Glycogen)_n$ + P_i $\longrightarrow$ ~~glucose 1-phosphate~~ + $(glycogen)_{n-1}$
~~Glucose 1-phosphate~~ + H_2O $\longrightarrow$ glucose + P_i

Net: $(Glycogen)_n$ + H_2O $\longrightarrow$ $(glycogen)_{n-1}$ + glucose

ATP COST

2 ATPs are required to store each glucose as glycogen.

Synthesis:[1]

Glucose + ATP $\longrightarrow$ ~~glucose 6-phosphate~~ + ADP
~~Glucose 6-phosphate~~ $\longrightarrow$ ~~glucose 1-phosphate~~
~~Glucose 1-phosphate~~ + ~~UTP~~ $\longrightarrow$ ~~UDP-glucose~~ + $2P_i$
~~UDP-glucose~~ + $(glycogen)_n$ $\longrightarrow$ ~~UDP~~ + $(glycogen)_{n+1}$
~~UDP~~ + ATP $\longrightarrow$ ~~UTP~~ + ADP

Net: $(Glycogen)_n$ + glucose + 2ATP $\longrightarrow$
                          $(glycogen)_{n+1}$ + 2ADP + $2P_i$

MOLECULAR FEATURES

Glycogen is a branched polymer (1-4 and 1-6 connections) of glucose connected in an α linkage at the anomeric carbon.

[1]To get the net reaction, molecules that occur on the right side of one reaction and on the left side of another reaction can be canceled (crossed through).

If there's plenty of glucose 6-phosphate around, there's no need to make more, so it might as well be stored.

The branched structure of glycogen poses some special problems for the synthesis and degradation of the molecule and for remembering how it's done (Fig. 9-2). Glycogen is a polymer of glucose in which linear strings of glucose molecules connected at the ends (through the 1 and 4 carbon atoms of the glucose) are strung together in a branched fashion. Branches occur where one glucose in the chain that's already connected 1-4 has another glucose attached at the 6 position. Another linear string of glucoses (attached 1-4) then takes off from the branch. This type of structure has directions, just like DNA. At one end (called the reducing end),[2] you have a glucose with nothing attached to carbon 1. Since each branch creates an extra end, glycogen has lots of ends that have nothing attached to carbon 4. The glucose with things attached at carbons 1, 4, and 6 is called a *branch point*. Special enzymes, branchers and de-branchers, are involved in making and destroying the branch points. Like much else in biology, these enzymes take what would appear to be a relatively simple task and complicate it beyond belief.

The degradation of glycogen is accomplished by the combined action of phosphorylase and glycogen debrancher.[3] Phosphorylase can make glucose 6-phosphate only out of unbranched glucose residues that are connected to glycogen in a 1-4 linkage. If the glucose has a branch on it, phosphorylase won't touch it. Phosphorylase cleaves the glycosidic bond of the glucose residues at the multiple, nonreducing ends and nibbles down the outer limbs of the glycogen molecule, releasing glucose 1-phosphates as it goes, until it gets to a structure that has 4 glucose molecules attached to each side of the branch. Then debrancher takes over. Debrancher takes 3 glucose residues from one side (C-6) of the branch and attaches them in a 1-4 linkage to the other side of the branch, leaving a structure in which a lone glucose is attached to the branch on the 1-6 side. The other side is now linear but is 7 glucoses long. The other activity of debrancher (yes, it has two activities in the same molecule) then takes the glucose off the 1-6 side and releases it as free glucose.[4]

---

[2] The *reducing end* is basically the end that doesn't have another glucose residue attached at carbon 1 (the anomeric carbon). It's called the reducing end because sugars that don't have anything attached at C-1 can be easily oxidized by specific chemicals that change colors, and such reactions fascinated early sugar chemists. If the end becomes oxidized, it must have reduced something . . . hence, the reducing end.

[3] Debrancher is given a terrible name in many texts—something like *gluc* something or other, or *glyco*, and maybe *transferase* stuck in there somewhere (actually it's amylo-1,6-glucosidase/4—glucanotransferase). You'll probably recognize it when you see it.

[4] Some free glucose (not glucose 1-phosphate) is released from glycogen, even in muscle. So the idea that muscle can't make any glucose is not quite right. However, this glucose is not really enough to count on.

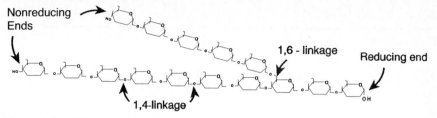

Figure 9-2   Glycogen Structure
Branches are created by forming glycosidic linkages with both the 4- and 6-hydroxyl groups of the glucose residue at the branch point. The glycogen polymer is very large and contains multiple branches.

You're left with a linear molecule (at least at this branch point), and phosphorylase is off and running again.

The synthesis of glycogen gets so complicated, it's hovering somewhere around 22 on the trivia sorter. Glycogen synthase adds a glucose from a UDP-glucose[5] to the C-4 end of the preexisting glycogen molecule. To put in branch points, the branching enzyme takes a block of 7 glucoses and transfers them to a site closer to the interior of the glycogen molecule . . . if the block of residues contains a free C-4 end, if it is contained in a block that's at least 11 long, and if the new branch point is at least 4 glucoses away from another branch. Got it?

---

[5] UTP + glucose $\longrightarrow$ UDP–O–glucose + PP$_i$. The oxygen from C-1 of glucose is attached to the UDP. The pyrophosphate is hydrolyzed to 2 P$_i$ by pyrophosphatase to drive the reaction to completion.

TCA CYCLE

·

· · · · · · · · · · · ·

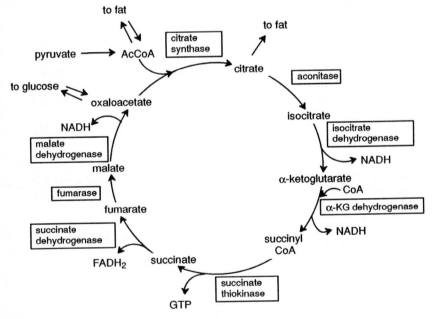

Figure 10-1   The Tricarboxylic Acid (TCA) Cycle

TCA CYCLE[1]

Function: To burn the acetyl-CoA made from fat, glucose, or protein in order to make ATP in cooperation with oxidative phosphorylation.

Location: All cells with mitochondria.

Connections: *From glycolysis* through acetyl-CoA.

Pyruvate makes oxaloacetate and malate through the anaplerotic reactions.

To β oxidation through acetyl-CoA.

To amino acid degradation through acetyl-CoA and various intermediates of the cycle.

Regulation: Supply and demand of TCA cycle.

Availability of NAD^+ and FAD as substrates.

Inhibition by NADH.

High-energy signals turn *off*.

Low-energy signals turn *on*.

ATP yield: Pyruvate $\longrightarrow$ 15ATP

Acetyl-CoA $\longrightarrow$ 12ATP

Equations:

$$Pyruvate + GDP + P_i + 3NAD^+ + FAD \longrightarrow$$
$$3CO_2 + GTP + 3NADH + FADH_2 + 3H^+$$

$$Acetyl\text{-}CoA + GDP + P_i + 2NAD^+ + FAD \longrightarrow$$
$$2CO_2 + GTP + 2NADH + FADH_2 + 2H^+$$

[1] The tricarboxylic acid cycle is also known as the Krebs cycle or the citric acid cycle. Why give something so central to life only one name?

FAT SYNTHESIS AND DEGRADATION

·

Fatty Acid Synthesis Function

Fatty Acid Synthesis Location

Fatty Acid Synthesis Connections

Fatty Acid Synthesis Regulation

Fatty Acid Synthesis ATP Costs (for C_{16})

Fatty Acid Synthesis Equation

Elongation and Desaturation

Triglyceride and Phospholipid Synthesis

β-Oxidation Function

β-Oxidation Location

Carnitine Shuttle

β-Oxidation Connections

β-Oxidation Regulation

β-Oxidation ATP Yield

β-Oxidation Equation

β-Oxidation of Unsaturated Fatty Acids

β-Oxidation of Odd-Chain-Length Fatty Acids

· · · · · · · · · · · ·

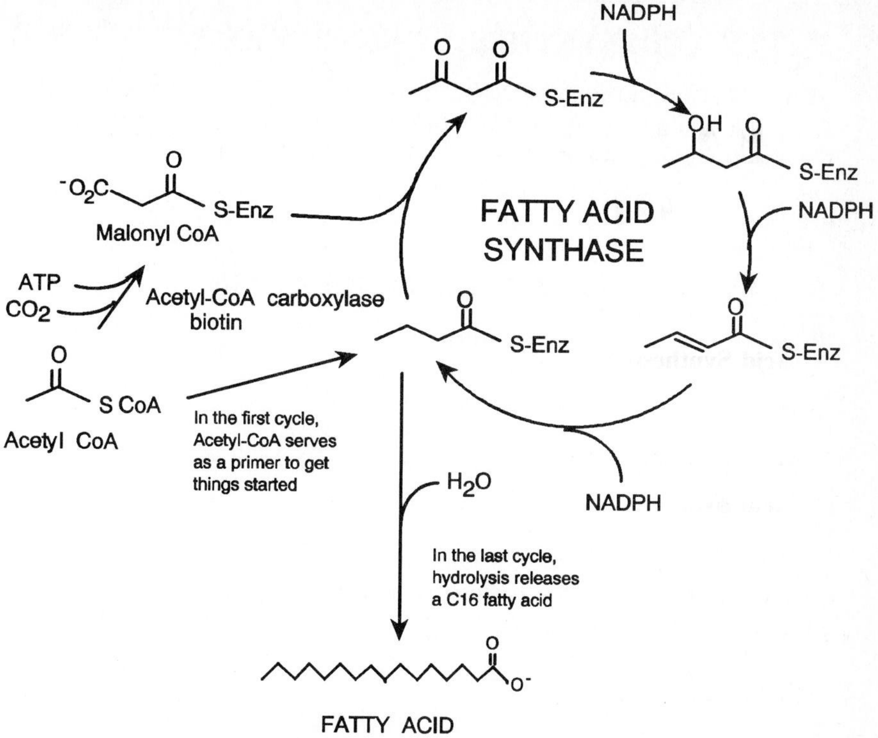

Figure 11-1 Fatty Acid Synthesis
The first formation of a carbon–carbon bond occurs between malonyl and acetyl units bound to fatty acid synthase. After reduction, dehydration, and further reduction, the acyl enzyme is condensed with more malonyl-CoA and the cycle is repeated until the acyl chain grows to C_{16}. When the growing fatty acid reaches a chain length of 16 carbons, the acyl group is hydrolyzed to give the free fatty acid.

FATTY ACID SYNTHESIS FUNCTION

To synthesize fatty acids from acetyl-CoA

FATTY ACID SYNTHESIS LOCATION

Liver and adipose cytoplasm

FATTY ACID SYNTHESIS CONNECTIONS

From *TCA* and mitochondria through *citrate* to C_{16} *fatty acid*
Through C_{16}-acyl-CoA to *elongation* and *desaturation* pathways
Through C_{16}-acyl-CoA to *triglycerides*

FATTY ACID SYNTHESIS REGULATION

Primary signals: *Insulin* turns *on.*
                          Glucagon turns *off.*
                          Epinephrine turns *off.*
                          Phosphorylation turns *off.*
Secondary signals: *Citrate* activates (acetyl-CoA carboxylase).

The major control point for fatty acid synthesis is acetyl-CoA carboxylase. The enzyme is inactivated by phosphorylation and activated by high concentrations of citrate.

FATTY ACID SYNTHESIS ATP COSTS (FOR C_{16})

8 acetyl-CoA$_{mito}$ $\longrightarrow$ 8 acetyl-CoA$_{cyto}$ -16 ATP
7 acetyl-CoA $\longrightarrow$ 7 malonyl-CoA -7 ATP
14NADPH $\longrightarrow$ 14NADP$^+$ -42 ATP
                              Total cost: -65 ATP

Calculating energy costs for the synthesis of a C_{16} fatty acid from acetyl-CoA is not as simple as you might first think. The major complication is that acetyl-CoA is made in the mitochondria, but fatty acid synthesis occurs in the cytosol—acetyl-CoA can't cross the mitochondrial membrane. Acetyl-CoA gets out of the mitochondria disguised as citrate. The acetyl-CoA is condensed with oxaloacetate to give citrate, and the citrate leaves the mitochondria. In the cytosol, the citrate is cleaved by an ATP-dependent citrate lyase into acetyl-CoA and oxaloacetate:

Citrate + ATP + CoA $\longrightarrow$ acetyl-CoA + ADP + P$_i$ + oxaloacetate

This reaction and the reactions required to get oxaloacetate back into the mitochondria set up the cycle shown in Fig. 11-2.

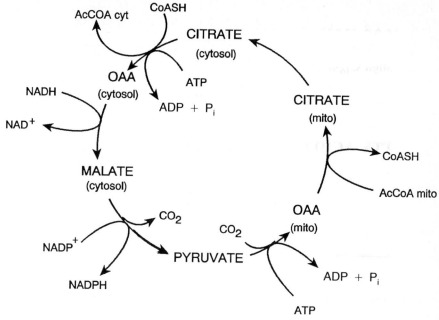

Figure 11-2
Getting **ACETYL-CoA OUT OF THE MITOCHONDRIA** and into the cytosol for
fat synthesis.

Another point that you should appreciate is that in cycles like this,
writing a balanced equation for the reaction is terribly easy. You just
write the stuff that goes into the cycle on the left and the stuff that comes
out on the right. It's not necessary to write down all the individual
reactions that make up the cycle itself. The compounds that are members
of the cycle itself (like oxaloacetate or citrate above) will not show up in
the final balanced equation. The balanced equation for Fig. 11-2 is

$$\text{Acyl-CoA}_{\text{mito}} + 2\text{ATP} + \text{CoA} + \text{NADH} + \text{NADP}^+ + \text{CO}_2 \longrightarrow$$
$$\text{CoA} + \text{acyl-CoA}_{\text{cyt}} + 2\text{ADP} + 2\text{P}_i + \text{NAD}^+ + \text{NADPH} + \text{CO}_2$$

If we cancel the things that appear on both the left and right, we have

$$\text{Acyl-CoA}_{\text{mito}} + 2\text{ATP} + \text{NADH} + \text{NADP}^+ \longrightarrow$$
$$\text{acyl-CoA}_{\text{cyt}} + 2\text{ADP} + 2\text{P}_i + \text{NAD}^+ + \text{NADPH}$$

In addition to moving acetyl-CoA from the mitochondria to the
cytoplasm, this cycle also converts an NADH to an NADPH. If we
assume that the amount of ATP that we could get from NADH and

NADPH oxidation is the same, making NADPH from NADH and $NADP^+$ doesn't cost any energy. So we can conclude that the cost of just moving the acetyl-CoA out of the mitochondria is 2 ATPs per acetyl-CoA.

The synthesis of C_{16} fatty acid from acetyl-CoA requires 1 acetyl-CoA and 7 malonyl-CoA. The synthesis of each malonyl-CoA requires an ATP (and the cofactor biotin).

$$\text{Acetyl-CoA} + CO_2 + \text{ATP} \longrightarrow \text{malonyl-CoA} + \text{ADP} + P_i$$

From here it's just the reaction catalyzed by fatty acid synthase.

$$\text{Acetyl-CoA} + 7 \text{ malonyl-CoA} + 14\text{NADPH} \longrightarrow$$
$$C_{16} \text{ fat} + 14\text{NADP} + 7CO_2 + 8\text{CoA}$$

If we count the NADPH (cytosol) as 3 ATP equivalents, which could have been oxidized by the TCA cycle if they hadn't been used for fatty acid synthesis, then the synthesis of C_{16} fat requires

$$
\begin{array}{lll}
8 \text{ acyl-CoA}_{mito} \longrightarrow 8 \text{ acyl-CoA}_{cyto} & -16 \text{ ATP} \\
7 \text{ acyl-CoA} \longrightarrow 7 \text{ malonyl-CoA} & -7 \text{ ATP} \\
14\text{NADPH} \longrightarrow 14\text{NADP}^+ & -42 \text{ ATP} \\
\hspace{3cm} \text{Total cost:} & -65 \text{ ATP}
\end{array}
$$

This is not the kind of number you want to remember, but if you're into ATP counting (and who isn't these days), you might want to understand how to go about figuring it out if you need to.

FATTY ACID SYNTHESIS EQUATION

$$\text{Acetyl-CoA} + 7 \text{ Malonyl-CoA} + 14\text{NADPH} + 14H^+ \longrightarrow$$
$$C_{16} \text{ fatty acid} + 14\text{NADP}^+ + 7CO_2 + 8\text{CoA}$$

Requires a phosphopantetheine cofactor

The reactions of fatty acid synthesis all occur on one enzyme—fatty acid synthase.[1] This enzyme has multiple catalytic activities on one polypeptide chain. The intermediates of the reaction are not released until the final C_{16} chain is completed. Each cycle of elongation adds 2 carbons

---

[1] A *synthase* is an enzyme that makes something but doesn't directly require the hydrolysis of ATP to do it. A *synthetase* requires the hydrolysis of ATP to make the reaction go.

to the growing chain. When the length gets up to 16 carbons, the fatty acid is released. In mammals, the product of fatty acid synthase is the free fatty acid, not the acyl-CoA.

• REACTIONS OF FATTY ACID SYNTHESIS:

Elongation by C_2:
$$R—C(=O)—S—Cys^2 + Mal—S—Pant^3 \longrightarrow$$
$$R—C(=O)CH_2—C(=O)—S—Pant + CoA + CO_2$$
Reduction:
$$R—C(=O)—CH_2—C(=O)—S—Pant + NADPH \longrightarrow$$
$$R—CH(OH)—CH_2—C(=O)—S—Pant + NADP^+$$
Dehydration:
$$R—CH(OH)—CH_2—C(=O)—S—Pant \longrightarrow$$
$$R—CH=CH—C(=O)—S—Pant + H_2O$$
Reduction:
$$R—CH=CH—C(=O)—S—Pant + NADPH \longrightarrow$$
$$R—CH_2—CH_2—C(=O)—S—Pant + NADP^+$$

Go to first step over and over until length is C_{16}.

Hydrolysis:
$$R—CH_2—CH_2—C(=O)—S—Pant + H_2O \longrightarrow$$
$$R—CH_2—CH_2—CO_2^- + E—Pant—SH$$

ELONGATION AND DESATURATION

Elongated by 2 carbon atoms at a time.
Cis double bonds can be introduced no farther than C-9 from the carboxylate end in mammals.
Numbers between double bonds are different by 3.
Δ Nomenclature—counts from *carboxylate* end.
ω Nomenclature—counts from *methyl* end.

---

[2] In the first cycle, this is acetyl-CoA. In subsequent cycles, the R group grows by two carbons each cycle.
[3] There's a lot of shuffling of acyl groups between the pantetheine thiol and the thiol of a cysteine residue of the enzyme. They're shown in the correct position for all the reactions, so you have to do an acyl transfer at the end of each cycle to put the growing acyl chain back on the cysteine residues. The elongation step takes place with the growing acyl chain on cysteine and the malonyl-CoA on the pantetheine. At the end of the condensation reaction, the elongated chain is on the pantetheine.

Once fatty acids have been made 16 carbons long, they can be lengthened by adding 2 carbon atoms at a time with malonyl-CoA in a reaction that looks a lot like the first step of fatty acid synthesis. However, the elongation reaction is carried out on the fatty acyl-CoA and by an enzyme that is different from fatty acid synthase.[4]

The inability of animals to put in double bonds at positions farther than 9 carbons from the carboxylate carbon atom (it's numbered 1) makes for obvious exam questions about whether a given unsaturated fatty acid comes from a plant or animal source. Animals can still elongate fatty acids even after they have been desaturated, so you can't just look at the number of the position of the double bond and decide whether or not animals could have made it. For example, if we take a C_{18} fatty acid with a double bond at C-9 (the first carbon of the double bond that you encounter when walking from the carboxylate end is at C-9), it can be elongated by adding 2 carbons to the carboxylate end. This changes the number of the double bond from C-9 to C-11. Each elongation step increases the number of each double bond by 2. However, if we count the number of carbons from the CH_3 end of the fatty acid, you'll notice the number isn't changed by elongating the fatty acid.

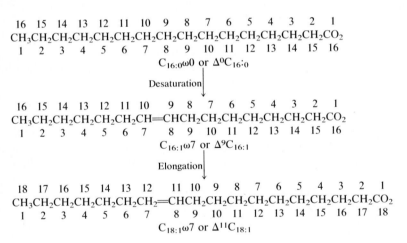

---

[4] Notice that fatty acid synthesis makes the fatty acid rather than the fatty acyl-CoA as a product. It would be a chemically simple task to take the fatty acid off the fatty acid synthase by a reaction with CoA. In fact, *Escherichia coli* does it this way and makes fatty acyl-CoA as a product. Most reactions that use the fatty acid (like desaturation, elongation, or triglyceride synthesis) require the fatty acyl-CoA, so we've got to turn right around and spend some ATP to make it when we could have just made it in the first place without spending any ATP. This strategy was obviously designed by a government committee.

Notice that elongation doesn't change the number of carbon atoms between the double bond and the CH_3 group at the left end. If a double bond is closer than 7 carbon atoms to the CH_3 group (numbering the CH_3 as 1, the first double bond you bump into would start at carbon 7), a plant must have made it.

Two nomenclature camps have grown up around the naming of unsaturated fatty acids. The first camp calls C-1 the carboxyl group and numbers the double bonds from this end. This is the delta (Δ) nomenclature. The number of carbons is given, followed by the number of double bonds. The positions of the double bonds from the carboxylate carbon are indicated as a superscript to the delta. The other camp, the omega (ω) system, numbers the chain from the CH_3 end. The position of the first double bond is indicated after the ω. Often in the ω system, the other double bonds are not specified except in the $C_{22:3}$ part. For example, $C_{22:3}$ $\omega 6$ would mean a C_{22} fatty acid with three double bonds, the first of which was 6 carbons from the CH_3 end. The other two double bonds would each be 3 carbons farther away from the CH_3 group (at 9 and 12 carbons from the CH_3 end). The ω nomenclature has the advantage that the position numbers of the double bonds do not change with elongation since the new carbons are added to the carboxylate end. In the Δ nomenclature, the $C_{22:3}\omega 6$ would be called $\Delta^{10,13,16}$ $C_{22:3}$.

ω 1 2 3 4 5 6═7 8 9═10 11 12═13 14 15 16 17 18 19 20 21 22

Δ 22 21 20 19 18 17═16 15 14═13 12 11═10 9 8 7 6 5 4 3 2 1

If you ever have to interconvert between these two nomenclatures, it may be easier for you to write down a representation of the structure rather than to try to figure out the relationships between the length, number of double bonds, and how you add and subtract what to get whatever.

TRIGLYCERIDE AND PHOSPHOLIPID SYNTHESIS

Glycerol phosphate comes from glycerol (*not in adipose*) or from dihydroxyacetone phosphate (in *liver* and *adipose*).
Nitrogen-containing phospholipids are made from diglyceride. Other phospholipids are made from phosphatidic acid.

(See Fig. 11-3.)

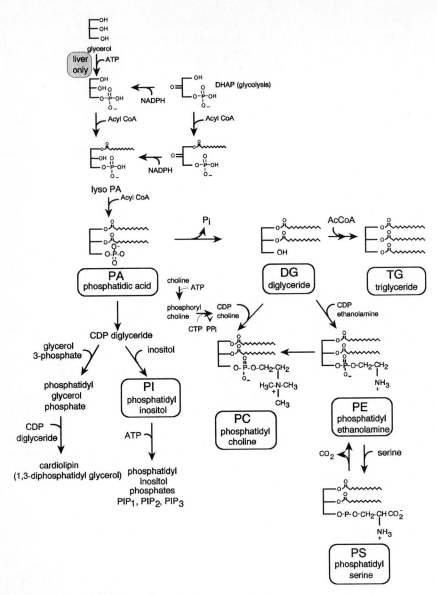

Figure 11-3   Synthesis of Phospholipids and Triglycerides
Glycerol utilization depends on the tissue. Adipose tissue can't use glycerol.
Nitrogen-containing phospholipids are made from diglycerides, while other phospholipids are made from phosphatidic acid (PA). PI = phosphatidylinositol; PC = phosphatidylcholine; PE = phosphatidylethanolamine; PS = phosphatidylserine.

β-OXIDATION FUNCTION

To break down fatty acids to acetyl-CoA to provide fuel for the TCA
   cycle

(See Fig. 11-4.)

β-OXIDATION LOCATION

Mitochondria of all tissues

CARNITINE SHUTTLE

Transfers fatty acids from cytoplasm to mitochondria for β oxidation
Inhibited by malonyl-CoA

Long-chain fatty acids can slowly cross the mitochondrial membrane
by themselves, but this is too slow to keep up with their metabolism. The
carnitine shuttle provides a transport mechanism and allows control of β
oxidation. Malonyl-CoA, a precursor for fatty acid synthesis, inhibits the
carnitine shuttle and slows down β oxidation (Fig. 11-5).

β-OXIDATION CONNECTIONS

Fatty acyl-CoA *in,* acetyl-CoA and NADH, FADH$_2$ *out*
From triglycerides through hormone-sensitive lipase

β-OXIDATION REGULATION

Primary signals (effects on hormone-sensitive lipase):
   Insulin turns *off.*
   Glucagon turns *on.*
   Epinephrine turns *on.*
   Phosphorylation turns *on.*
Secondary signals: Malonyl-CoA inhibits carnitine acyltransferase.

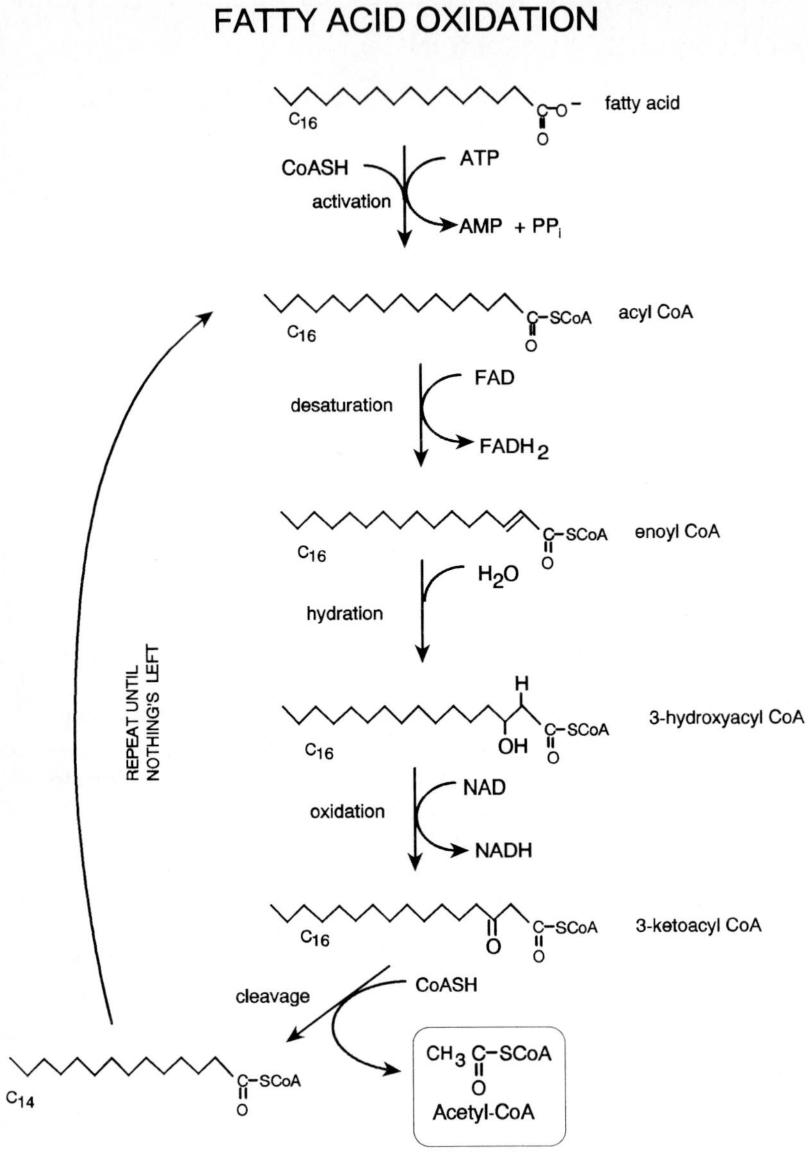

FATTY ACID OXIDATION

Figure 11-4   Fatty Acid Oxidation (β oxidation)

The major hormone-sensitive control point for the mobilization of fat and the β-oxidation pathway is the effect of phosphorylation on the activity of the hormone-sensitive lipase of the adipose tissue. The major direct control point for β oxidation is the inhibition of carnitine

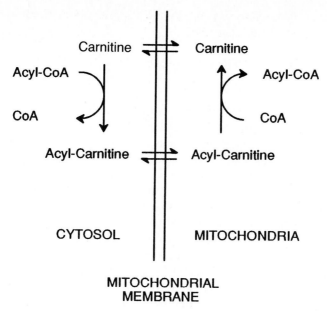

Figure 11-5
The **CARNITINE SHUTTLE** is used to transport fatty acids into the mitochondria.

acyltransferase by malonyl-CoA. Since the malonyl-CoA required for fatty acid synthesis inhibits β oxidation, this regulation keeps the opposed pathways of fatty acid synthesis and β oxidation in check. You don't do synthesis and degradation at the same time.

β-OXIDATION ATP YIELD

C_{16} fatty acid $\longrightarrow$ $16CO_2$	129 ATP
C_{16} fatty acyl-CoA $\longrightarrow$ $16CO_2$ + CoA	131 ATP

Breakdown into steps:

Activation of fatty acid to fatty acyl-CoA	-2 ATP
7 $FADH_2$ made from forming double bond at C-2 (7×2)	14 ATP
7 NADH made from oxidations during formation of 3-ketoacyl-CoA (7×3)	21 ATP
8 acetyl-CoA (8×12) through TCA cycle	96 ATP

When calculating ATP yields from β oxidation, you have to be careful to notice whether you start with the fatty acid or with the fatty acyl-CoA. Two high-energy phosphate equivalents are required to activate fatty acids to the acyl-CoA.

Fatty acid—CO_2^- + ATP + CoASH $\longrightarrow$
$$\text{fatty acyl—C(=O)—SCoA + AMP + PP}_i$$

All the ATP comes from oxidative phosphorylation coupled to the metabolism of acetyl-CoA by the TCA cycle. No oxygen, no β oxidation.

Each cycle of β oxidation reduces the length of the fatty acid chain by 2 carbons, produces 1 acetyl-CoA (12 ATP), 1 $FADH_2$ (2 ATP), and 1 NADH (3 ATP). Each 2-carbon unit of the fatty acid then results in the production of 17 ATPs. To figure out how much ATP a C_{20} fatty acid could make, you might remember that C_{16} fatty acid makes 129 and then add 17 × 2 for the additional 4 carbons (129 + 34 = 163). That seems like the hard way. It may be easier just to remember that each acetyl-CoA can give 12 ATPs and that each cycle generates 1 $FADH_2$ and 1 NADH. A C_{20} fatty acid would make 10 acetyl-CoA, 9 $FADH_2$, and 9 NADH and require 2 ATP equivalents for activation [(12 × 10) + (9 × 2) + (9 × 3) − 2 = 163 ATP]. Notice that breaking a C_{20} fatty acid into 10 acetyl-CoA units requires only nine β-oxidation cycles—the last cycle gives 2 acetyl-CoA as the product.

β-OXIDATION EQUATION

C_{16} fatty acid + CoA + ATP $\longrightarrow$ C_{16}-acyl-CoA + AMP + PP_i

C_{16}-acyl-CoA + 7NAD^+ + 7FAD $\longrightarrow$
$$\text{8 acetyl-CoA + 7NADH + 7FADH}_2 + 7\text{H}^+$$

Each reaction of β oxidation is catalyzed by a different enzyme. Chemically, they're pretty much the same as the reverse of the individual reaction of fatty acid synthesis, with two exceptions: (1) β oxidation uses FAD for the formation of the double bond at the C-2 position, and (2) the reactions occur with the fatty acid attached to CoA rather than to the pantetheine of a multienzyme complex.

• INDIVIDUAL REACTIONS OF β OXIDATION:

Activation:
R—CH_2—CH_2—CO_2^- + CoA + ATP $\longrightarrow$
$$\text{R—CH}_2\text{—CH}_2\text{—C(=O)—SCoA + AMP + PP}_i$$

Oxidation:
$$R—CH_2—CH_2—C(=O)—SCoA + FAD \longrightarrow$$
$$R—CH=CH—C(=O)—SCoA + FADH_2$$
Hydration:
$$R—CH=CH—C(=O)—SCoA + H_2O \longrightarrow$$
$$R—CH(OH)—CH_2—C(=O)—SCoA$$
Oxidation:
$$R—CH(OH)—CH_2—C(=O)—SCoA + NAD^+ \longrightarrow$$
$$R—C(=O)—CH_2—C(=O)—SCoA + NADH$$
Cleavage:
$$R—C(=O)—CH_2—C(=O)—SCoA + CoA \longrightarrow$$
$$R—C(=O)—SCoA + CH_3C(=O)—SCoA$$

What β oxidation actually accomplishes is the removal of a C-2 unit as acetyl-CoA from the carboxyl end of the fatty acid. This keeps happening until the fatty acid is completely converted to acetyl-CoA.

β OXIDATION OF UNSATURATED FATTY ACIDS

Double bond initially on *odd* carbon (Δ system):
  Isomerize *cis*-3 C=C to *trans*-2 C=C then proceed with
    normal β oxidation.
  2 fewer ATPs per double bond
Double bond initially on *even* carbon (Δ system):
  Hydrate *cis*-2 C=C to D-3-hydroxyacyl-CoA, epimerize D-3-
    hydroxy to L and continue normal β oxidation.
  2 fewer ATPs per double bond
  Or reduce 2-*trans*-4-*cis* C=C to 3-*trans* with NADPH; then
    isomerize the 3-*trans* to 2-*trans* and proceed as with normal β
    oxidation.
  5 fewer ATPs per double bond

If a fatty acid already has a double bond in it, the scheme by which the fatty acid is oxidized depends on where the double bond ends up after several of the C-2 fragments have been removed by normal β oxidation. With a double bond already present, the enzyme that catalyzes the first step (insertion of the double bond at C-2) gets confused when there is already a double bond at C-2 or at C-3. The fact that the double bonds in unsaturated fatty acids are invariably *cis* also complicates life since the double bond introduced at C-2 by the desaturating enzyme of β oxidation is a *trans* double bond.

As the β-oxidation machinery chews off 2-carbon fragments, it nibbles down to one of two possible situations depending on whether the first double bond started out at an *even-* or an *odd*-numbered carbon when counting from the carboxylate end. If the double bond is on an odd-numbered carbon (as in *cis* $\Delta^9 C_{18:1}$), it is metabolized slightly differently than a fatty acid in which the unsaturation is on an even-numbered carbon (as in *cis* $\Delta^{12} C_{18:1}$).

If the double bond is on an odd carbon, β oxidation removes 2-carbon fragments until it gets to the structure with a 3-*cis* double bond [R–CH=CH–CH$_2$–C(=O)–SCoA]. A new double bond can't be placed between C-2 and C-3 because there's already a double bond at C-3. In this situation, the activity of an isomerase simply moves the double bond from C-3 to C-2 and at the same time makes sure that the configuration is *trans*. From this point on, the metabolism is just like normal β oxidation (hydration, oxidation, cleavage). If you're counting ATPs, these unsaturated fatty acids produce 2 fewer ATPs for each double bond since there is no FADH$_2$ produced by putting in the double bond (see Fig. 11-6).

If the double bond starts out at an even carbon, β oxidation runs its normal course until the structure *cis*-2-R–CH=CH–C(=O)–SCoA is reached. The rub here is that the double bond is on an OK carbon (C-2), but it's in the wrong configuration. The double bonds in unsaturated fatty acids are invariably of the *cis* configuration, but β oxidation introduces the double bond at C-2 in the *trans* configuration. The pathway cited in most texts involves the addition of water to the 2-*cis* double bond to give the 3-hydroxy species just as in normal β oxidation, except for a mean twist. If the double bond is introduced in the *trans* configuration by β oxidation itself, hydration gives the L-3-hydroxy fatty acyl-CoA. But if the double bond is in the *cis* configuration, hydration gives the D-3-hydroxy fatty acyl-CoA. The configuration around C-3 (D vs. L) might appear trivial to you, but to the enzyme that oxidizes the C-3 (C–OH) to the carbonyl (C=O), it's night and day. The dehydrogenase won't touch the D configuration because the OH group is in the wrong place relative to the R and CoA groups. To get around this problem, there's an enzyme (an epimerase) that converts the D to the L epimer. The L epimer is then recognized by the dehydrogenase, and it's smooth sailing from there on. Since the isomerase and epimerase don't require ATP hydrolysis, ATP counting through this pathway would show that a double bond at an *even* position reduces the yield of ATP by 2 (no FADH$_2$ is formed in the first desaturation reaction).

A somewhat newer pathway, which may be the real pathway in many cells, has been discovered recently. This may or may not be described in your text. In this pathway, the fatty acyl-CoA is metabolized normally, introducing a *cis* double bond at C-2 and removing 2-carbon fragments

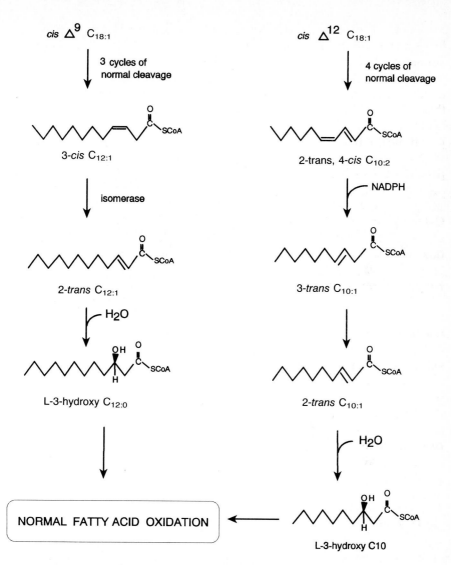

Figure 11-6   Metabolism of Unsaturated Fatty Acids
This may not be the pathway in your text. If you've not seen this mentioned,
ignore it. The other (and possibly incorrect) pathway is simpler anyway.

until the fatty acid contains two double bonds, *2-trans-4-cis-*
R–CH=CH–CH=CH–C(=O)–S–CoA. The 2-*trans* double bond is put in
by the normal β oxidation, and the 4-*cis* is left from the original *cis* double
bond in the unsaturated fatty acid. At this point, the 2-*trans*-4-*cis* fatty
acyl-CoA is reduced by an NADPH-dependent reductase to the *trans*-3-

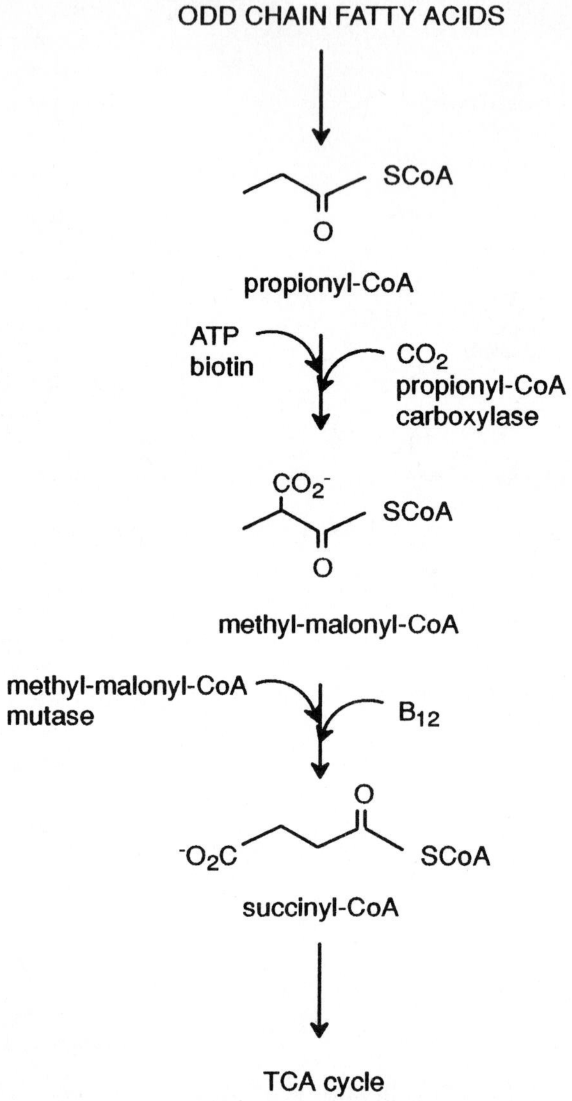

Figure 11-7
Metabolism of **ODD-CHAIN-LENGTH FATTY ACIDS** yields propionyl-CoA, which can be rearranged to succinyl-CoA and dumped into the TCA cycle.

R–CH$_2$–CH=CH–CH$_2$–C(=O)–S–CoA. The 3-*trans* double bond is then isomerized to the 2-*trans* double bond, and β oxidation proceeds as normal. If you count ATPs by this pathway, each double bond at an even carbon would decrease the yield of ATP by 5 (assuming that the NADPH used is equivalent to NADH and would produce 3 ATPs if oxidized by the electron transport chain and that no FADH$_2$ is made in putting in the double bond).

β OXIDATION OF ODD-CHAIN-LENGTH FATTY ACIDS

Makes propionyl-CoA, which is metabolized by propionyl-CoA carboxylase (biotin) and methylmalonyl-CoA mutase (B$_{12}$) to give succinyl-CoA.

Since there's no good way to make a C-1 fragment during β oxidation, the metabolism of fatty acids with an odd number of carbons must finally give you a 3-carbon piece as propionyl-CoA. Odd-chain-length fatty acids aren't very abundant in nature (some shellfish and bacteria make odd-chain-length fatty acids), but they may be somewhat more common on exams. The propionyl-CoA resulting from the β oxidation of odd-chain fatty acids is metabolized by a weird pathway that is also used to metabolize the propionyl-CoA produced by the breakdown of the amino acid threonine. Two vitamins are required in this pathway, biotin and B$_{12}$. This is one of two places you'll see vitamin B$_{12}$ (the other is in the metabolism of 1-carbon fragments) (see Fig. 11-7).

ELECTRON TRANSPORT AND OXIDATIVE PHOSPHORYLATION

·

Oxidation and Reduction

The Electron Transport Chain

Connections

Regulation

P/O Ratios

Uncouplers

Inhibitors

· · · · · · · · · · · ·

You only have to look at the ATP yield from the TCA cycle, 12 of them per molecule of acetyl-CoA, to know that oxidative phosphorylation must be important. That's where all the electrons from NADH and $FADH_2$ go after they're made by the TCA cycle.

OXIDATION AND REDUCTION

Oxidation is the loss of electrons.
  NADH is oxidized to NAD^+.
Reduction is the gain of electrons.
  O_2 is reduced to H_2O.

Electrons usually aren't floating around in space; they're usually stuck on some atom or other. The simple consequence of this is that when one thing loses electrons, something else must gain them. Every oxidation of something must be coupled to the reduction of something else. The molecule or atom that loses the electrons has been oxidized; the one that gains them has been reduced. Oxidants, or oxidizing agents, are compounds that oxidize other compounds—they are reduced in the process. Reductants, or reducing agents, are compounds that reduce other compounds—they are oxidized in the process.

$$CH_3CCO_2^- + NADH + H^+ \rightleftharpoons CH_3CHCO_2^- + NAD^+$$

$$\underset{O}{\|} \qquad\qquad\qquad \underset{OH}{|}$$

$$\text{Pyruvate} \qquad\qquad\qquad\qquad \text{Lactate}$$

Pyruvate is reduced to lactate. Lactate is oxidized to pyruvate. NADH is oxidized to NAD^+. NAD^+ is reduced to NADH. Pyruvate and NAD^+ are oxidizing agents. Lactate and NADH are reducing agents.

As fuel molecules are oxidized, the electrons they have lost are used to make NADH and $FADH_2$. The function of the electron transport chain and oxidative phosphorylation is to take electrons from these molecules and transfer them to oxygen, making ATP in the process.

THE ELECTRON TRANSPORT CHAIN

Two electrons flowing down the chain make
   3 ATP/NADH
   2 ATP/$FADH_2$

As electrons move down the electron transport chain, the carriers become reduced (Fig. 12-1). The next carrier oxidizes the previous carrier, taking its electrons and transferring them on to the next carrier. Finally the electrons end up reducing oxygen to water. The cytochromes are named with letters in no particular order, making them tough to memorize, but you probably should learn them, at least right before the exam—after that you can look them up if you ever need to.

The energetics of the electron transport steps makes the process work. Overall there's a lot of free energy lost in the transfer of electrons from NADH to oxygen—the overall reaction is very favorable, with an equilibrium constant that's overwhelmingly large. At the three sites

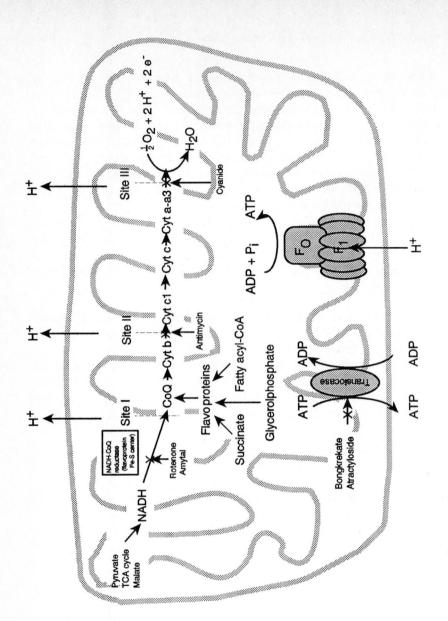

Figure 12-1 The Electron Transport Chain

where ATPs are made (labeled I, II, and III), the reaction is the most downhill.

During the electron transfers at the three classic sites of phosphorylation (marked I, II, and III), protons are pumped out of the mitochondria into the cytoplasm. The exact number of protons pumped at each site is somewhat controversial; however, this proton pumping makes the interior of the mitochondria alkaline.

ATP is made by the F_1F_0 ATPase. This enzyme allows the protons back into the mitochondria. Since the interior is alkaline, the reaction is favorable—favorable enough to drive the synthesis of ATP by letting protons back into the mitochondria. Exactly how the F_1F_0 ATPase couples the flow of protons down their concentration gradient to the formation of ATP is not known in molecular detail. The proton flow through the F_1F_0 ATPase is required to release ATP from the active site where it was synthesized from ADP and P_i. The ATP is made in the interior of the mitochondria and must be exchanged for ADP outside the mitochondria to keep the cytosol supplied with ATP. The exchange of mitochondrial ATP for cytoplasmic ADP is catalyzed by the ATP/ADP translocase.

The complete transfer of 2 electrons from NADH through the entire electron transport chain to oxygen generates 3 ATPs. $FADH_2$ feeds electrons into coenzyme Q (a quinone) after the first ATP-generating step. Flavin-linked substrates (those that make $FADH_2$) generate only 2 ATPs per 2 electrons transferred down the chain. Flavin-linked substrates generate less ATP, not only because they feed in after the first ATP has already been made; they make 2 ATPs because $FADH_2$ is not as strong a reducing agent as NADH. There is not enough energy in the oxidation of $FADH_2$ to generate 3 ATPs.

CONNECTIONS

NADH and $FADH_2$ from the TCA Cycle.
Electrons from NADH outside the mitochondria are transported into the mitochondria by the malate-aspartate shuttle or the α-glycerol phosphate shuttle.
O_2 is a gas supplied by the blood.
ADP outside the mitochondria is swapped for ATP inside the mitochondria by a specific translocase.
F_1F_0 ATPase couples H^+ gradient to ATP synthesis.

The electron transport chain gets its substrates from the NADH and $FADH_2$ supplied by the TCA cycle. Since the TCA cycle and electron transport are both mitochondrial, the NADH generated by the TCA cycle can feed directly into oxidative phosphorylation. NADH that is generated outside the mitochondria (for example, in aerobic glycolysis) is not transported directly into the mitochondria and oxidized—that would be too easy.

There are two shuttles involved in getting the electrons from NADH into mitochondria. The α-glycerol phosphate shuttle works most simply. In this shuttle, NADH in the cytoplasm is used to reduce dihydroxyacetone phosphate (DHAP) to α-glycerol phosphate. The α-glycerol phosphate is actually the molecule transported into the mitochondrion, where it is oxidized back to DHAP, giving mitochondrial $FADH_2$. The DHAP then leaves the mitochondrion to complete the shuttle. With this shuttle in operation, there's a cost. Normally, the oxidation of mitochondrial NADH gives 3 ATPs. However, the mitochondrial enzyme that oxidizes α-glycerol phosphate uses FAD as the oxidizing agent. The $FADH_2$ that results gives only 2 ATP equivalents. Using this shuttle, the cytoplasmic NADH yields only 2 ATPs.

The other shuttle is the malate-aspartate shuttle. The advantage of this shuttle is that it gives you 3 ATPs for the oxidation of each cytoplasmic NADH. In red muscle, heart, and brain tissues the malate-aspartate shuttle is the major pathway for shuttling electrons into mitochondria. In white muscle, the α-glycerol phosphate shuttle predominates (Fig. 12-2).

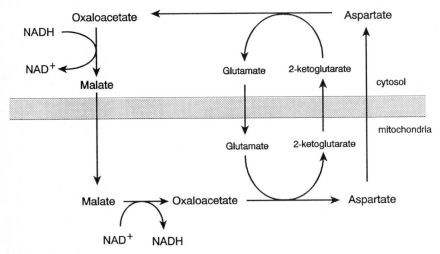

Figure 12-2
The **MALATE-ASPARTATE SHUTTLE** gets reducing equivalents (electrons) from cytosolic NADH into the mitochondria so that 3 ATPs can be made.

REGULATION

The rate of oxidative phosphorylation is controlled by the supply of ADP and phosphate.

Assuming that oxygen is available and that there is a supply of NADH- or $FADH_2$-generating substrates, the activity of oxidative phosphorylation is determined by the availability of ADP. If ADP is available and there is enough phosphate around (there usually is), the ADP and P_i are converted to ATP. If not, not.

P/O RATIOS

These are the numbers of ATP equivalents made per 2 electrons passed down the electron transport chain.

NADPH: P/O = 3
$FADH_2$: P/O = 2
Succinate (with rotenone present): P/O = 2
Acetate: P/O = 2.5

The P/O ratio is the number of ATPs made for each O atom consumed by mitochondrial respiration. The P stands for high-energy-phosphate equivalents, and the O actually stands for the number of $\frac{1}{2} O_2$'s that are consumed by the electron transport chain. The full reduction of O_2 to $2 H_2O$ takes 4 electrons. Therefore, 2 electrons reduce $\frac{1}{2}$ of an O_2. The oxidation of NADH to NAD and the oxidation of $FADH_2$ to FAD are both 2-electron oxidations. O can be read as the transfer of 2 electrons. It's not quite as obscure as it sounds.[1]

To figure out a P/O ratio[2] you have to figure out two things—the P and the O. The P is easy if you've learned to count ATPs made by various metabolic pathways. The P is the net number of ATPs made by the metabolism of the substance you're dealing with. The O is a little harder.

---

[1] It probably *is* as obscure as it sounds. Remember O = transfer of 2 electrons down the entire chain.

[2] Pronounced "pee to oh," "pee over oh," or "pee oh"—all are used.

Here you must figure out how many times 2 electrons have been passed down the electron transport chain. For each NADH or each $FADH_2$ made during the metabolism of your substance, 1 O is consumed as 2 electrons are passed down the chain.

In the presence of the inhibitor rotenone (to prevent the oxidation of NADH by the electron transport chain), succinate can be metabolized only to fumarate, producing an $FADH_2$ in the process.

$$Succinate + FAD \longrightarrow fumarate + FADH_2$$

The oxidation of the $FADH_2$ makes 2 ATPs and consumes 1 O. The P/O for succinate is 2.

In the absence of rotenone, the NADH that is made from the conversion of succinate to oxaloacetate can be oxidized by the electron transport chain. The metabolism of succinate then becomes

$$Succinate + FAD \longrightarrow \text{fumarate} + FADH_2$$

$$\text{Fumarate} + H_2O \longrightarrow \text{malate}$$
$$\text{Malate} + NAD^+ \longrightarrow oxaloacetate + NADH + H^+$$

Net: $Succinate + FAD + H_2O + NAD^+ \longrightarrow$
$$oxaloacetate + FADH_2 + NADH + H^+$$

In this case, succinate metabolism to oxaloacetate produces 5 ATPs from the oxidation of the NADH and $FADH_2$—2 from the $FADH_2$ and 3 from the NADH. Two O's are consumed, 1 for each of the NADH and $FADH_2$ molecules. The P/O is then $5/2 = 2.5$. Thus the P/O can be a nonintegral number.

Just one more. Let's do acetate. Before it can be metabolized, acetate must be activated in a reaction that uses 2 ATP equivalents.[3]

$$Acetate + CoA + ATP \longrightarrow acetyl\text{-}CoA + ADP + PP_i$$

Acetyl-CoA metabolized through the TCA cycle yields 3 NADH, 1 $FADH_2$, and 1 GTP—a total of 12 ATP equivalents (3 from each NADH, 2 from each $FADH_2$, and 1 GTP—12 in all). Four O's are used, 1 for each NADH and $FADH_2$. The P in this case is 10 ($12 - 2$ for the activation of acetate to acetyl-CoA). The $P/O = 10/4 = 2.5$. P/O ratios for anything else are calculated in the same way.

---

[3] For reactions that make PP_i (pyrophosphate), the PP_i is rapidly hydrolyzed to 2 P_i in the cell, so we'll consider the formation of PP_i to use 2 high-energy-phosphate bonds.

UNCOUPLERS

Allow protons back into the mitochondria without making any ATP
Stimulate oxygen consumption

Mitochondria do three things: oxidize substrates, consume oxygen, and make ATP. Uncouplers prevent the synthesis of ATP but do not inhibit oxygen consumption or substrate oxidation. Uncouplers work by destroying the pH gradient. The classic uncoupler is dinitrophenol (DNP). This phenol is a relatively strong acid and exists as the phenol and the phenolate anion.

$$2,4\text{-DNP}\text{—OH} + H_2O \rightleftharpoons 2,4\text{-DNP}\text{—O}^- + H_3O^+$$

Because both the anion and the acid are lipophilic (greasy) enough to cross the mitochondrial membrane, DNP can transport protons across the membrane and destroy the pH gradient. The 2,4-DNP–OH crosses from the cytosol into the mitochondrion, carrying its proton with it. In the more alkaline environment of the mitochondrion, the DNP–OH loses its proton and the pH falls. The 2,4-DNP–O$^-$ then leaves the mitochondrion and repeats the cycle again until the pH inside is the same as the pH outside. With no pH gradient, there is no ATP synthesis. However, there is still oxidation of substrates and consumption of oxygen. With no ATP synthesis, the ADP concentration is high and the electron transport chain keeps trying to pump out protons. In fact, uncouplers usually stimulate oxygen and substrate consumption. Long-chain fatty acids can uncouple mitochondria by the same mechanism. There are other ways to collapse the pH gradient. Valinomycin is a potassium ionophore[4] and collapses the electrochemical gradient[5] of the mitochondria. Collapsing the electrochemical gradient also prevents ATP synthesis.

---

[4] An *ionophore* is a compound that is capable of selectively carrying ions across a membrane. The ion fits into a specific binding site in a molecule that is hydrophobic enough to cross the membrane. There are calcium-specific ionophores, proton ionophores, sodium ionophores et cetera.

[5] *Electrochemical gradient* is the name given to the gradient of charge and ions that exists across the inside and the outside of a cellular membrane. The outside of the mitochondrion is more positively charged than the inside, and the concentration of potassium ions is higher outside than inside. When an ion falls through (an expression that gives a nice image, but *moves* would be just as good) the electrochemical gradient, it is driven by both its concentration gradient and the charge difference between the inside and the outside of the membrane. Ions tend to move from areas of high concentration to areas of low concentration. Positive ions tend to move toward the more negative compartment. The bigger the charge difference between the inside and the outside, the bigger the free-energy difference that drives the ion movement.

INHIBITORS

Inhibitors block the flow of electrons at a specific site and inhibit
    electron flow and ATP synthesis.
Inhibitors inhibit oxygen consumption and ATP synthesis.

Inhibitors actually block one of the steps of oxidative phosphorylation. *Cyanide* blocks the last step of electron transfer by combining with
and inhibiting cytochrome oxidase. The effect is just like oxygen deprivation. The less obvious effect is that all the electron carriers become more
reduced than without the inhibitor. The reason is that the substrates are
still pushing reducing equivalents (electrons) down the electron transport
chain. But it's blocked at the end. The result is that all the carriers before
the block become reduced. For the same reason, inhibiting electron
transport also tends to keep the NADH and $FADH_2$ reduced (depending
on where the inhibitor acts). Carriers after the block become more oxidized. Carries after the block can still transfer their electrons to oxygen.
Once they've done this, though, there are no more reducing equivalents
available because of the block, and they are left in the oxidized state.

Different inhibitors block at different points of the chain. The general
rule is that all electron carriers that occur before the block become
reduced and all that occur after the block become oxidized.

Rotenone inhibits the transfer of electrons from NADH into the
electron transport chain. The oxidation of substrates that generate NADH
is, therefore, blocked. However, substrates that are oxidized to generate
$FADH_2$ (like succinate or α-glycerol phosphate) can still be oxidized and
still generate ATP. Because NADH oxidation is blocked, the NADH pool
becomes more reduced in the presence of rotenone since there's nowhere
to transfer the electrons.

Atractyloside and bongkrekate inhibit the entry of ADP into the
mitochondria. After all the ADP in the mitochondria has been converted
to ATP, oxidative phosphorylation stops, since the ATP that's made can't
get out and new ADP can't get in from the outside.

INHIBITOR	SITE	EFFECT
Cyanide	Cytochrome oxidase	Blocks transfer of electrons to O_2. Blocks at site III.
Antimycin	Electron transfer from cyt b to cyt c_1	All intermediates before and including cyt a will be in the reduced state; all intermediates after and including cyt c_1 will be in the oxidized state. Blocks at site II.
Rotenone	NADH-CoQ reductase	Blocks oxidation of NADH (site I). NADH will become reduced; Substrates like succinate that enter via FADH will still be oxidized and make 2 ATPs/mol.
Oligomycin	ADP phosphorylation	Blocks phosphorylation of ADP. Does not inhibit uncoupled oxidations.
Atractyloside and bongkrekate	ADP-ATP transporter	Inhibits entry of ADP into mitochondria and ATP export. Stops electron transport because of lack of ADP. Inside, all ADP is converted to ATP.

PENTOSE PHOSPHATE PATHWAY

·

PENTOSE PHOSPHATE PATHWAY

Function:	To supply reducing equivalents for biosynthesis (NADPH) and pentoses for DNA and RNA biosynthesis.
Location:	Everywhere.
Connections:	To *glycolysis* and *glycogen* through *glucose 6-phosphate*.
	Shares some enzymes with glycolysis.
	To *DNA-RNA synthesis* through *ribose 5-phosphate*.
Regulation:	NADPH inhibits.
	NADP$^+$ activates.
	High glucose 6-phosphate activates.

EQUATIONS:

Making NADPH:

3 Glucose 6-P + 6NADP$^+$ $\longrightarrow$
2 fructose 6-P + glyceraldehyde 3-P + 3CO$_2$ + 6NADPH + 6H$^+$

Making ribose 5-P:

Glucose 6-P + 2NADP$^+$ $\longrightarrow$
$\qquad\qquad\qquad$ ribose 5-P + CO$_2$ + 2NADPH + 2H$^+$

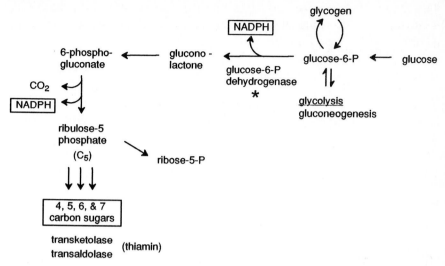

Figure 13-1   The Pentose Phosphate Pathway

The hexose monophosphate pathway has several names just to confuse you. It's called the hexose monophosphate shunt or pathway (HMP shunt or pathway), or the pentose phosphate pathway, or the phosphogluconate pathway. The pathway in its full form is complicated and has complicated stoichiometry. Usually it's not necessary to remember all of it. The important points are that it makes NADPH for biosynthesis and riboses (C-5 sugars) for DNA and RNA synthesis.

NADPH is a reducing agent that is reserved for biosynthetic pathways—notably fatty acid synthesis. Thus, the HMP pathway is called upon when reducing equivalents and fatty acid synthesis are turned on. Primarily, the regulation of the pathway is through the supply and demand of NADPH.

NADPH is also used to keep the cellular (and mitochondrial) glutathione in the reduced form through the action of glutathione reductase:[1]

$$GSSG + NADPH \longrightarrow 2GSH + NADP^+$$

---

[1] Glutathione (GSH) is a tripeptide, γ-glutamylcysteinylglycine. The γ-glutamyl part means that the amino group of the cysteine is attached to the side chain carboxyl group of the glutamic acid rather than to the α-carboxyl. GSH is a thiol (sulfhydryl)-containing molecule that can be oxidized to the disulfide, GSSG, by cellular oxidants such as hydrogen peroxide. It helps protect cells against damage that can be caused by oxidants.

This reaction also protects proteins with cysteine residues from becoming oxidized to the disulfide since the GSH can be used to reduce the protein disulfide back to the thiol form:

$$\text{Protein—SS—Protein} + 2\text{GSH} \rightleftharpoons 2 \text{ protein—SH} + \text{GSSG}$$

A somewhat more trivial thing to remember about the HMP pathway is that this is one of the places you've seen the vitamin thiamin pyrophosphate. This cofactor is necessary for the transketolase reaction that is in the middle of the HMP pathway. The transketolase reaction converts two C-5 sugars to a C-7 and a C-3. The other place you've seen thiamin pyrophosphate as a cofactor is in the pyruvate dehydrogenase and α-ketoglutarate dehydrogenase reactions.

AMINO ACID METABOLISM

·

Nonessential Amino Acid Synthesis

Essential Amino Acids

Amino Acid Degradation

Products of Amino Acid Degradation

Generalities of Amino Acid Catabolism

· · · · · · · · · · · ·

NONESSENTIAL AMINO ACID SYNTHESIS

AMINO ACID	SYNTHETIC ROUTE
Ala	from Pyruvate by transamination
Glu	from α-Ketoglutarate by transamination
Asp	from Oxaloacetate by transamination
Gln	Glu + NH_4^+ + ATP → Gln
Asn	Asp + Gln + ATP → Asn + AMP + PP_i + Glu
Ser	Glucose → hydroxypyruvate → Ser
	Glucose → phosphohydroxypyruvate → Ser
Gly	Ser + THfolate → Gly + CH_2-THfolate
Arg	Glu → Glu-semialdehyde → ornithine → Arg
Pro	Glu → Glu-semialdehyde → Pro
Tyr	Phe → Tyr (phenylalanine hydroxylase, biopterin cofactor)
Cys	Met → homoCys + Ser → cystathionine → Cys

The other *nine* amino acids are essential and must be taken from the diet. Notice that some of the amino acids require other amino acids for their synthesis. Exam questions usually center on whether or not an amino acid is essential and the metabolites that serve as precursors for specific amino acids.

ESSENTIAL AMINO ACIDS

His, Ile, Leu, Lys, Met, Phe, Thr, Trp, Val

AMINO ACID DEGRADATION

Ketogenic:	Leu, Lys Degraded to acetyl-CoA. Glucose cannot be made from these.
Glucogenic and Ketogenic:	Ile, Phe, Tyr, Trp Goes both ways.
Glucogenic:	*Everything else* Degraded to pyruvate or a member of the TCA cycle Glucose can be made from these.

The complete catabolic pathways of the individual amino acids are a complex set of pathways that are probably not worth remembering in detail (this is obviously opinion). This doesn't mean they're not important. In fact, there are diseases that are caused by inherited defects in most of the pathways. The table above is a general guide that shows where the amino acids go and points out significant intermediates.

GENERALITIES OF AMINO ACID CATABOLISM

If a vitamin or cofactor is involved in amino acid metabolism, it's most likely pyridoxal phosphate (B_6), unless it involves serine, and then it's B_6 and folic acid.

Nitrogen is dumped into the urea cycle by transamination to make Asp or Glu or by deamination to make ammonia.

PRODUCTS OF AMINO ACID DEGRADATION

Ala to *pyruvate* by transamination
Arg to urea and *glutamate*
Asp to *oxaloacetate* by transamination or to fumarate via urea cycle
Asn to Asp
Cys carbon to *pyruvate,* sulfur to *sulfate*
Glu to α-*ketoglutarate* by transamination, then to glucose
Gln to *glutamate* by hydrolysis
Gly to *glyoxylate* or *serine*
His to *glutamate* and *one-carbon pool*
Met to *propionyl-CoA* via homocysteine → cystathionine →
    ketobutyrate
Pro to *glutamate*
Ser to *glycine* and *CH₂THfolate*
Thr to *propionyl-CoA* through ketobutyrate
Val to *propionyl-CoA* through transamination, decarboxylation, and
    a bunch of rearrangements
Leu to *acetoacetate* and *acetyl-CoA* through transaminaton, decar-
    boxylation, and a bunch of rearrangements
Ile to *propionyl-CoA* through transamination, decarboxylation, and
    a bunch of rearrangements
Phe to *Tyr,* then to *acetoacetate* and *fumarate*
Tyr to *acetoacetate* and *fumarate*
Trp to *acetyl-CoA* via ring oxidation and cleavage to ketoadipate
Lys to *acetyl-CoA* via transamination and deamination to ketoadi-
    pate

The nitrogen contained in the amino acids is usually disposed of through the urea cycle. One of the early, if not the first, steps in amino acid catabolism involves a transamination using oxaloacetate or α-ketoglutarate as the amino-group acceptor. This converts the amino acid into a 2-keto acid, which can then be metabolized further.

$$R—CH(NH_3^+)CO_2^- + \text{oxaloacetate} \rightleftharpoons R—(C{=}O)CO_2^- + Asp$$

These enzymes invariably involve a cofactor, pyridoxal phosphate (vitamin B_6). In addition, pyridoxal phosphate is also required for most decarboxylations, racemizations, or elimination reactions in which an amino acid is a substrate. Pyridoxal phosphate is not involved in decar-boxylations in which the substrate is not an amino acid. So if a question asks something about an amino acid and a vitamin, the answer will most likely be pyridoxal phosphate. There are a couple of exceptions in which pyridoxal phosphate may not be the answer to a vitamins–amino acid question. If the amino acid is serine, then the answer might also include folic acid (the reaction here is the conversion of serine to glycine with the formation of methylene tetrahydrofolic acid–see the section in Chap. 19 on one-carbon metabolism). The other place you might see a vitamin other than pyridoxal phosphate is in the metabolism of propionyl-CoA, a product of the catabolism of some amino acids. In this case, the vitamin may be B_{12} (the conversion of methylmalonyl-CoA to succinyl-CoA—see Odd Chain Fatty Acid Oxidation in Chap. 11).

The nitrogen from the amino groups of most amino acids is transami-nated into glutamate or aspartate at some point in the degradative scheme. This nitrogen enters the urea cycle as glutamate, which is reduc-tively deaminated by glutamate dehydrogenase to yield ammonia or by the reaction of aspartate with citrulline to give argininosuccinate (urea cycle).

· C H A P T E R · 15 ·

INTEGRATION OF ENERGY METABOLISM

Integrating Metabolic Pathways

ATP

Glucose

Storage Molecules

Metabolic States and Signals

Insulin

Glucagon

Epinephrine

Secondary Signals

Generalities of Metabolism

Phosphorylation

Glycogen

Metabolic Movements of Glycogen

Fat

Metabolic Movements of Fat

Protein

Metabolic Movements of Protein

Tissue Cooperation

Liver

Muscle

Adipose

Brain

Connection of Storage Pools

Feeding

Fasting

Starvation

Excitement

Interorgan Cycles

Cori Cycle

Alanine Cycle

Ketone bodies

Welcome to energy metabolism. It's the nuts and bolts of life at the cellular level.[1] Energy metabolism is responsible for maintaining a constant supply of ATP in all different tissues. Since some tissues, like red cells and brain, rely heavily on glucose to make ATP, maintaining the ATP levels in all tissues requires maintaining the availability of glucose. The purpose of all the pathways we'll discuss is to maintain ATP and glucose supplies.

As you prowl around the individual metabolic pathways of energy metabolism, you need to get a feel for the kinds of things you should be looking for as you examine each metabolic scheme. There are four individual pathways involved in energy metabolism: glycolysis-gluconeogenesis, fatty acid synthesis–β oxidation, glycogen synthesis-degradation, and the TCA cycle. First look for the overall function of each pathway: What does it do? This involves knowing the molecules coming into the pathway and those going out. Next, understand when the pathway should be on or off—what metabolic states require the pathway to function. Then, figure out which tissues use the pathway and how. Finally, see how the behavior of the pathway may be integrated into the cooperation between organs and tissues.

Energy metabolism makes sense if you realize that each individual pathway and each organ has a function. Understanding metabolism in every detail may well be impossible, but understanding the general themes is not only possible but important.

Energy metabolism maintains the supply of ATP and glucose by making storage molecules (*glycogen, fat, protein*) when food is available and by retrieving ATP and glucose from storage when they are needed. The need for glucose or ATP may constitute a demand for massive amounts of immediate energy or it may simply be the need to maintain energy and glucose levels between meals. Energy metabolism is regulated in a manner that involves extensive cooperation between different organs, but the goal is always the same: to maintain adequate ATP and glucose levels.

INTEGRATING METABOLIC PATHWAYS

Metabolic pathways interconnect *glycogen, fat,* and *protein* reserves to store and retrieve ATP and glucose.

Most of energy metabolism should make sense—don't forget that. The function of energy metabolism is reasonably simple—*all* of it is

[1]At least the nuts part is right.

concerned with putting energy into storage when it's not needed and taking energy out of storage when it is needed. It's the details that complicate the picture.

There are only three types of storage molecules: *glycogen, fat,* and *protein.* These three storage forms are all connected, and that's what energy metabolism is all about—the connections. When metabolism works normally, we're assured of a relatively constant supply of energy and glucose.

No one tissue can survive metabolically without the others. Each of the four major tissue types (liver, muscle, adipose, brain) has a specialized metabolic function. There are some differences in the metabolic pathways in each tissue; however, these differences are relatively simple and serve to specialize the metabolic functions of the different tissue types. There is real cooperation between the different organs. Each organ has its own metabolic profile, its own needs, and its own capabilities.

ATP

ATP is the immediate source of energy for cellular processes.

Ultimately, the energy source for all cellular processes is the hydrolysis of ATP. There are only two major sources of ATP—the TCA cycle–electron transport, and glycolysis. The major source of ATP is the mitochondrial electron transport chain fueled by the TCA cycle. Since the mitochondrial electron transport requires oxygen, much of our ATP production is directly linked to the supply of oxygen. Red cells, since they don't have any mitochondria, must rely totally on anaerobic glycolysis for energy, and muscle may be forced to rely only on anaerobic glycolysis when strenuous exercise uses up oxygen faster than it can be delivered to the muscle.

GLUCOSE

Glucose is essential for the metabolism of fat and for providing ATP
    in red cells and brain.
Glucose cannot be made from fat.
Anaplerotic reactions keep the TCA cycle turning.
    Pyruvate carboxylase
    Malic enzyme

Glucose (or more correctly its metabolites) is essential for the functioning of the TCA cycle. For the TCA cycle to keep turning, the intermediates of the cycle must be maintained at a reasonable level. Since these intermediates are used for things other than the TCA cycle, they must be replaced constantly. The trouble is that the intermediates of the TCA cycle cannot be synthesized from fat (our most abundant storage form of energy). So we've got to have glucose or its equivalent to burn fat.

The reactions that convert pyruvate to intermediates of the TCA cycle are called the *anaplerotic reactions*. Pyruvate, which can be made only from glucose or some of the amino acids, can be converted to oxaloacetate by the enzyme pyruvate carboxylase or to malate by malic enzyme.

Pyruvate carboxylase (a biotin-dependent carboxylase):

$$CH_3C(\!\!=\!\!O)CO_2^- + ATP + CO_2 \longrightarrow {}^-O_2CCH_2C(\!\!=\!\!O)CO_2^- + ADP + P_i$$
Pyruvate   Oxaloacetate

Malic enzyme:

$$CH_3C(\!\!=\!\!O)CO_2^- + NADPH + CO_2 + H^+ \longrightarrow$$
Pyruvate

$${}^-O_2CCH_2CH(OH)CO_2^- + NADP^+$$
Malate

These reactions result in the net synthesis of TCA-cycle intermediates. They are necessary to replace TCA-cycle intermediates that are withdrawn from the cycle and used for other things.

Cells that do not have mitochondria (like red cells) must use glucose for energy since they have no TCA cycle or oxidative phosphorylation. Without a constant glucose supply, these cells would die. Other tissues like brain rely heavily on glucose metabolism for energy; however, brain can adapt to use alternative energy sources if glucose is not available.

STORAGE MOLECULES

Glycogen: Glucose storage
Fat: ATP storage
Protein: ATP and glucose storage

The way energy metabolism is set up, there are actually two things that must be stored—glucose and energy. Glycogen is a branched polymer of glucose that is accumulated in liver, kidney, and muscle as a short-term glucose supply. The glycogen stores of muscle, because of the amount of muscle we have, are the largest stores in terms of mass. However, the muscle stores glycogen only for its own needs. Because muscle is missing the enzyme glucose 6-phosphatase, it cannot convert glycogen (or anything else) into glucose. The liver and kidney do have glucose 6-phosphatase activity and share their glycogen stores to help maintain glucose levels throughout the body.

Fat is the major way we store energy. Fat is stored principally in adipose tissue and can be stored in virtually unlimited amounts. Fat is metabolized through β oxidation to acetyl-CoA and then through the TCA cycle, where it's burned completely to CO_2 to make ATP. Fat cannot be used to make carbohydrate because acetyl-CoA *cannot* be converted directly to precursors of glucose without losing its carbon atoms first (you'll see what this really means later).

Protein molecules make up the structural and functional elements of cells; however, they can be used to provide energy. Proteins are undergoing a constant cycle of synthesis and degradation. In times of need, the protein mass of the body can be used to generate both energy and glucose. The amino acids derived from protein breakdown can be used for energy or the production of glucose equivalents. Protein is a storage form for both glucose and ATP.

METABOLIC STATES AND SIGNALS

STATE	RESPONSE	SIGNAL
Feeding:	*Store fat, glucose, protein*	High insulin, low glucagon
Fasting:	*Retrieve glucose and fat*	Low inuslin, high glucagon
Starvation:	*Retrieve fat and protein*	Low insulin, high glucagon
Diabetes:	*Retrieve fat and protein*	Low insulin, high glucagon
Excitement:	*Retrieve glucose and fat quickly*	High epinephrine

We'll distinguish four basic metabolic states. There are three major metabolic signals that we'll consider—insulin, glucagon, and epinephrine.

Feeding is eating, which we do several times a day. After feeding, the precursors of the storage molecules are in abundant supply. Some of the food intake is simply burned to supply immediate energy, but the general theme after eating is to store some of the food intake as glycogen, fat, and protein so that we can use it later.

After eating, blood glucose levels are high and the pancreas secretes insulin as a signal that tells the rest of the body that glucose can be had. Insulin is a signal that means high blood glucose levels. Insulin promotes the entry of glucose into insulin-sensitive cells.

Fasting is not eating. Hopefully, we fast several times a day, the longest one being overnight. During fasting, we withdraw some of the energy supplies we stored after eating. As blood glucose levels begin to fall because we haven't eaten lately, the levels of insulin begin to fall and the levels of glucagon, a hormone that signals low blood glucose levels, begin to rise. Glucagon promotes the retrieval of energy from all its storage forms.

Starvation is not eating for days—or longer. Humans can adapt to survive food deprivation for quite a long period of time—30 to 60 days. We store enough glycogen for about 24 hours without glucose intake. After this time, glycogen stores are depleted and we must turn to alternative sources for glucose equivalents—protein. A number of other adaptations occur that prolong the survival of some tissues, particularly brain.

Diabetes results from a lack of insulin secretion by the pancreas. Without insulin, cells take up glucose very slowly. The lack of insulin results in an inability to use blood glucose for fuel. Consequently, the body behaves as if it were starving even though food is available. The metabolic responses of the untreated insulin-dependent diabetic are essentially the metabolic responses of starvation.

Excitement is the rapid onset of an immediate need for energy. It can be considered a short-term, intense starvation. The adrenal medulla in response to an excitement signal (each of us has our own) dumps epinephrine into the circulation.

INSULIN

Insulin is a signal for high blood glucose levels.
It *stimulates synthesis* of glycogen, fat, and protein.
It *inhibits breakdown* of glycogen, fat, and protein.
It increases glucose transport into cells.

Insulin, secreted by the β cells of the pancreas in response to rising blood glucose levels, is a signal that glucose is abundant. It's saying, "Store it while it's here." Insulin binds to a specific receptor on the cell surface and exerts its metabolic effect by a mechanism that's not fully understood. Note that insulin stimulates storage processes and at the same time inhibits degradative pathways.

GLUCAGON

Glucagon is a signal for low blood glucose levels.
It *stimulates breakdown* of glycogen, fat, and protein.
It *inhibits synthesis* of glycogen, fat, and protein.
It *increases protein phosphorylation*.
  It activates cAMP-dependent protein kinase.

Glucagon, made by the α cells of the pancreas, is the antithesis of insulin. Its effects are just the opposite. Glucagon is a signal that says, "Remove things from storage." A good bit is known about how glucagon exerts its effect on cells. Glucagon increases the activity of specific cellular protein kinases (enzymes that use ATP to phosphorylate a serine or a threonine, and occasionally a tyrosine residue of some specific proteins).

When glucagon levels are high, specific proteins (we'll see which ones later) are phosphorylated—high glucagon levels mean increased phosphorylation. *Phosphorylation activates specific enzymes that need to be active when energy and glucose supplies are low. Conversely, phosphorylation inactivates enzymes that are involved in energy storage.*

Glucagon binds to a cell-surface receptor. When it is occupied, the receptor, through the intermediacy of a coupling protein (a G protein), activates adenylate cyclase. Activated adenylate cyclase takes ATP and makes cAMP and P_i from it. The cAMP then activates cAMP-dependent protein kinase by binding to the inactive enzyme and releasing an inhibitory subunit. The active protein kinase then goes about kinasing (phosphorylating) other proteins, some of which are kinases themselves. The net result is a huge amplification of the original signal—kinases activating kinases activating kinases—and increased phosphorylation of cellular proteins (Fig. 15-1).

The G protein serves as a timekeeper. The G protein binds GTP, and with GTP bound, it can couple the receptor and activate the adenylate cyclase. However, the G protein slowly hydrolyzes the bound GTP to GDP and P_i. When this happens, the whole complex falls apart and the adenylate kinase is inactivated.

When glucagon levels fall, cAMP phosphodiesterase destroys the accumulated cAMP, and specific protein phosphatases remove the phosphate from the phosphoproteins. These phosphatases themselves are often regulated by phosphorylation—yes, there are phosphatase kinases

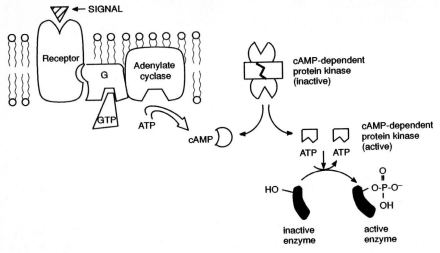

Figure 15-1
The **PROTEIN KINASE CASCADE** amplifies the original extracellular signal by increasing levels of cAMP, which activates cAMP-dependent protein kinase, which phosphorylates specific proteins.

and phosphatase phosphatases. It's real easy to lose it here, but the key factor is that increased glucagon levels lead to increased protein phosphorylation, and decreased glucagon levels lead to decreased protein phosphorylation.

EPINEPHRINE

Epinephrine is a signal that energy is needed immediately.
It *stimulates breakdown* of glycogen, fat, and protein.
It *inhibits synthesis* of glycogen, fat, and protein.

Epinephrine, a hormone made in the adrenal medulla and sympathetic nerve endings, calls for rapid mobilization of energy and glucose. Epinephrine, like glucagon, binds to specific cellular receptors and activates adenylate cyclase. For the most part, epinephrine can be considered to have an effect similar to that of glucagon. Epinephrine, too, leads to increased protein phosphorylation.

SECONDARY SIGNALS

HIGH ENERGY	LOW ENERGY	HIGH GLUCOSE	LOW GLUCOSE
ATP	cAMP	Fructose 2,6-P_2	cAMP
Citrate	AMP	Glucose 6-P	
Fatty acids	ADP		
NADH	P_i		
Acetyl-CoA			

The primary hormonal signals serve as extracellular signals that are interpreted by a signal transduction apparatus and turned into signals within the cell—these second messengers like cAMP and fructose 2,6-bisphosphate warn individual enzymes within the cell about what's happening outside.

The levels of other cellular molecules may also affect the activities of specific enzyme and metabolic pathways. These secondary signals can be grouped into four classes: high-energy signals, low-energy signals, high-glucose signals, and low-glucose signals. The consequence of a rise in the concentration of one of these metabolites is just what you would expect. Not all these molecules affect all enzymes and/or pathways. It's just that if they do have an effect, it will be in the direction indicated by the type of the signal. For example, fructose 2,6-bisphosphate is a signal for the presence of glucose. Therefore, you would expect that increased levels of fructose 2,6-bisphosphate would increase the metabolism of glucose through glycolysis, increase fatty acid synthesis, and increase the storage of glycogen and protein. However, all that fructose 2,6-bisphosphate does is to increase the activity of glycolysis and decrease gluconeogenesis by affecting the activities of phosphofructokinase and fructose 1,6-bisphosphatase. It doesn't directly affect the activities of other enzymes. Fructose 2,6-bisphosphate is a local signal.

There are two ways to go here, particularly with the effectors like citrate, ATP, AMP, P_i, and the like. You could memorize exactly which enzymes are affected by these secondary signals, or you could just realize that the signals should make sense. Rather than just memorize, try to decide what the effector should do—most of the time you'll be right. It's not necessary to know that ATP inhibits phosphofructokinase, the regulatory enzyme of glycolysis. Just reason that ATP is a high-energy signal; glycolysis makes energy; and ATP should inhibit phosphofructokinase—it does.

GENERALITIES OF METABOLISM

1. ATP and glucose levels must be reasonably constant.
2. The utilization of fat for energy requires carbohydrate.
3. Glucose cannot be made from fat.
4. Synthesis and degradative pathways don't happen at same time.
5. Low energy levels turn on glycolysis and lipolysis.
6. Low glucose levels turn on gluconeogenesis and protein degradation.
7. Protein phosphorylation in response to rising cAMP levels activates enzymes involved in maintaining blood glucose levels and in making ATP.

There are a few generalities of metabolism that will help you understand metabolism and why a lot of things work the way they do.

• 1. ATP AND GLUCOSE MUST BE AVAILABLE AT ALL TIMES. Much of metabolism is involved with generating just enough ATP to balance the demands. When energy and glucose are plentiful, ATP and glucose are stored by the synthesis of fat, protein, and glycogen. When energy and glucose are needed, they are generated by converting the storage molecules to glucose and energy. The constant demand for ATP is obvious. It's always needed to drive energy-dependent processes like movement and thinking. The need for glucose or the need to consider it apart from its ability to generate ATP by direct metabolism may be less obvious. First, some tissues can use only glucose as an energy source, and their energy supply must be protected. Second, all tissues require glucose or an equivalent molecule to burn fat as an energy source (generality 2). When ATP and glucose supplies are abundant, energy and glucose equivalents are stored so that later they can be withdrawn and used when energy and glucose supplies are low.

• 2. GLUCOSE IS REQUIRED TO METABOLIZE FAT. While fat provides much ATP, it cannot be metabolized without some source of carbohydrate. As we'll see below, glucose (or the intermediates of the TCA cycle) cannot be made from acetyl-CoA. The only place we can get the TCA-cycle intermediates is from glucose or protein (amino acids). TCA-cycle intermediates are used for a variety of things (amino acid synthesis, purine synthesis, etc.) so that they are continuously being consumed and must be resupplied to keep the TCA cycle going. When glucose or its storage form, glycogen, is plentiful, the TCA-cycle intermediates malate and oxaloacetate can be made directly from pyruvate (the anaplerotic reactions). When glucose is not plentiful, the TCA-cycle intermediates must be derived from the degradation of amino acids (proteins). Not only do we have to have glucose to keep red cells going, but we have to have glucose to keep burning fat for energy.

• 3. GLUCOSE CANNOT BE MADE FROM FAT.   The end product of fat metabolism is acetyl-CoA. Acetyl-CoA cannot be used to generate glucose. There are arrows that you can follow on metabolic pathways that will get from acetyl-CoA to glucose, but you really can't get there with anything left. The problem is that to make the oxaloacetate needed for the synthesis of glucose from acetyl-CoA you need oxaloacetate. As we go around the TCA cycle back to oxaloacetate, two carbons are added from acetyl-CoA, but two carbon atoms are lost as CO_2 (not the same ones we added as acetyl-CoA). The result is that there can be no net synthesis of new oxaloacetate (or malate) from acetyl-CoA—basically because oxaloacetate synthesis from acetyl-CoA requires oxaloacetate. This means that metabolism of fat cannot supply glucose. Normally dietary intake and glycogen stores provide this supply. Under conditions of starvation, protein and the glycerol from triglyceride degradation serve as sources of glucose.

• 4. SYNTHESIS AND DEGRADATION DON'T HAPPEN AT THE SAME TIME.   Although they may share some common steps (such as in glycolysis and gluconeogenesis), synthetic and degradative pathways are not the simple reverse of each other. Synthetic pathways always use more ATP than you can make by the degradative pathway. If both synthetic and degradative pathways were on at the same time, the net result would be the hydrolysis of ATP (Fig. 15-2).

This "wasteful" hydrolysis of ATP has been termed a *futile cycle* since it apparently doesn't do anything. However, there may be some advantages to a few futile cycles. Some ATP hydrolysis may be used to generate heat. There may also be some control benefits to futile cycles. Keeping the engine running in a car is wasteful, but it allows you to respond rapidly to changes in situation. Keeping a futile cycle going can allow rapid changes in the flux through the cycle in either direction and, as a consequence, more sensitive control.

To minimize futile cycling, signals that turn on a synthetic pathway always turn off the opposing degradative pathway.

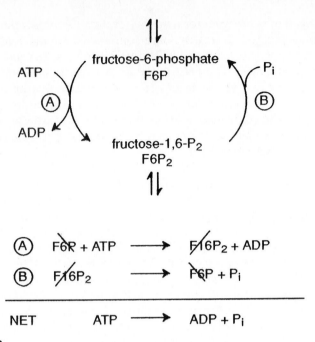

Figure 15-2
A **FUTILE CYCLE** is set up when both synthetic and degradative pathways are operating at the same time. The net reaction of a futile cycle is just ATP hydrolysis.

• 5. LOW ENERGY LEVELS TURN ON LIPOLYSIS AND GLYCOLYSIS. The TCA cycle coupled to mitochondrial oxidative phosphorylation is the major way of generating ATP. Acetyl-CoA destined for the TCA cycle can be generated by either the metabolism of glucose (glycolysis) or the degradation of fat (β oxidation). Low energy then mobilizes the stores of glycogen and lipids.

• 6. LOW GLUCOSE LEVELS TURN ON GLUCONEOGENESIS AND PROTEIN DEGRADATION. There are two ways to maintain blood glucose levels—degradation of glycogen and synthesis of glucose from pyruvate (gluconeogenesis). Glycogen is stored in the liver and muscle. Gluconeogenesis is the major domain of the liver and kidney. When the glucose level is low, liver and kidney supply glucose to the blood to maintain the supply by stimulating the degradation of glycogen and by stimulating the synthesis of glucose through gluconeogenesis. Glycogen has somewhat different functions in different tissues. In the liver and kidney, glycogen can be degraded to supply glucose to the rest of the body or it can be used for energy. In muscle, glycogen can only be used locally, for generating energy through glycolysis. The muscle glycogen stores cannot make significant free glucose because skeletal muscle is missing the enzyme glucose 6-phosphatase. Thus, what happens to glycogen depends on the tissue as well as the overall metabolic state.

• 7. PROTEIN PHOSPHORYLATION IN RESPONSE TO RISING cAMP LEVELS ACTIVATES ENZYMES THAT MAINTAIN GLUCOSE LEVELS AND RETRIEVE ENERGY. PHOSPHORYLATION INACTIVATES ENZYMES THAT STORE GLUCOSE, FAT, AND PROTEIN.

PHOSPHORYLATION

Low energy and low glucose levels are associated with an increased activity of cAMP-dependent protein kinase and *increased phosphorylation.*

Enzymes required only during *low-energy or low-glucose* situations are *activated by phosphorylation.*

Enzymes required only during *high-energy or high-glucose* situations are *inactivated by phosphorylation.*

A major way to control enzyme activity is the reversible phosphorylation of serine or threonine residues. It's such a major way to regulate enzymes that you will spend much of your time trying to remember whether or not enzyme X is activated or inactivated by phosphorylation, and you will invariably forget one or two of them on the exam. By remembering a couple of generalities,[2] you can actually figure out a lot of the effects of phosphorylation on specific enzymes without really memorizing them.

Protein phosphorylation is an ATP-dependent reaction catalyzed by numerous protein kinases:

$$\text{Protein—OH} + \text{ATP} \longrightarrow \text{Protein—O—PO}_3^{2-} + \text{ADP}$$

This modification is not directly reversible, but the phosphate group can be removed from the protein by the action of protein phosphatases.

$$\text{Protein—O—PO}_3^{2-} + \text{H}_2\text{O} \longrightarrow \text{Protein—OH} + \text{P}_i$$

[2] As with most generalities, there are invariably exceptions. If you have time to memorize the few exceptions, go ahead.

Phosphorylation activates some proteins and inactivates others. The actual phosphorylation of a regulated protein is often catalyzed by a protein kinase that is specific for one or just a few proteins; however, the protein kinases themselves are often regulated by phosphorylation-dephosphorylation mechanisms. At least for the regulatory enzymes of energy metabolism, the thing that usually starts the whole phosphorylation mess is cAMP-dependent protein kinase. This enzyme is activated by increases in the levels of cAMP, a second messenger for low energy and low glucose levels. Activation of cAMP-dependent protein kinase ultimately leads to increased protein phosphorylation.

For example, phosphorylation of phosphorylase, the enzyme responsible for degrading glycogen to glucose 1-phosphate, activates the enzyme. Glycogen degradation in both liver and muscle is required under low-glucose or low-energy conditions, conditions associated with increased protein phosphorylation. In contrast, glycogen synthase, which must be turned *off* under conditions of low glucose levels, is inactivated by phosphorylation.

The easy way to decide if phosphorylation activates or inactivates a given enzyme is to decide whether the pathway that uses the enzyme should be on or off under conditions favoring phosphorylation (low energy or low glucose). If the pathway should be on under low-energy or low-glucose conditions, phosphorylation should activate the pathway and the enzyme in question. If the pathway should be off, phosphorylation should inactivate the enzyme. Phosphorylase is simple enough, and the regulation of most other enzymes can be figured out using the principle outlined above. The example below, taken from the regulation of glycolysis and gluconeogenesis, is probably the most complicated one you'll see.

How does phosphorylation affect the activity of phosphofructo-2-kinase (PFK-2), the enzyme that synthesizes fructose 2,6-bisphosphate, a regulator of glycolysis? There are two possible answers: it either activates it or inactivates it. The simplest approach to the question is just to flip a coin. You should stand a 50:50 chance of getting it right. The next simplest way is to figure it out.

Fructose 2,6-bisphosphate stimulates glycolysis by allosterically activating phosphofructo-1-kinase (PFK-1).[3] First, decide what should happen to the overall pathway (glycolysis) when cAMP levels are high,

---

[3] It's unfortunate that we have to deal with PFK-1 and PFK-2. PFK-1 is the enzyme that catalyzes the formation of fructose 1,6-bisphospate from fructose 6-phosphate. PFK-2 makes the 2,6-bisphosphate.

because this will tell you what has to happen to the activity of the enzyme upon phosphorylation. If the activity of the overall pathway must be high, then phosphorylation should work to increase the activities of enzymes in the pathway and decrease the activities of enzymes in opposing pathways. The answer to this question may depend on the tissue, but the thinking process does not.

In liver, cAMP activates gluconeogenesis, but in muscle, it activates glycolysis. Let's do liver first, and the muscle answer will just be the opposite. So . . . we want to activate gluconeogenesis in liver in response to increased phosphorylation (increased levels of cAMP). Phosphorylation of our enzyme (PFK-2) must have an effect that is consistent with the activation of gluconeogenesis. If gluconeogenesis is on and glycolysis is off, the level of fructose 2,6-bisphosphate (an activator of glycolysis) must fall. If fructose 2,6-bisphosphate is to fall, the PFK-2 that synthesizes it must be made inactive. So, in liver, phosphorylation of PFK-2 must inactivate the enzyme.

In muscle, phosphorylation of PFK-2 in response to increased cAMP activates the enzyme, the level of fructose 2,6-bisphosphate rises, and glycolysis is activated.

There's also a fructose 2,6-bisphosphatase that hydrolyzes fructose 2,6-bisphosphate; see if you can figure out what happens to the phosphatase activity in liver and muscle when the enzyme is phosphorylated. As a check to your answer, you might recall that PFK-2 and fructose 2,6-bisphosphatase are one and the same protein. Phosphorylation-dephosphorylation actually shifts the activity of this single protein between the kinase and the phosphatase. So the answer you get should be opposite to the one we got above.

GLYCOGEN

Gylocgen is a short-term reserve of glucose.

Liver and Kidney: 24-hour supply to share with brain and red cells
Muscle: Stores glycogen for its own needs; lack of glucose 6-phosphatase prevents release of glucose for other tissues
Adipose: No significant glycogen stores
Red cells: No significant glycogen stores
Brain: No significant glycogen stores

METABOLIC MOVEMENTS OF *GLYCOGEN*

Feeding (↓ glucagon ↑ *insulin*):   *Glycogen synthesis*
Fasting (↑ *glucagon* ↓ insulin):   *Glycogen degradation*
Starvation (↑ *glucagon* ↓ insulin):   *Glycogen exhaustion*
Diabetes (↑ *glucagon* ↓ insulin):   *Glycogen degradation*
Excitement (↑ *epinephrine*):   *Glycogen degradation*

Glycogen serves two roles. In muscle, glycogen is an energy reserve that is degraded when muscle ATP demands are high. In liver and kidney, glycogen is a storage form of glucose and is specifically degraded when blood glucose levels drop (Fig. 15-3).

After eating, when glucose is abundant, liver, kidney, and muscle put glucose equivalents into storage for retrieval later. Between meals, when glucose is absent from the diet, liver and kidney break down glycogen to supply blood glucose. Muscle, on the other hand, is selfish. Skeletal muscle does not supply other organs with glucose from its glycogen; it uses it to supply glucose 6-phosphate and energy for itself. Muscle doesn't make free glucose because it doesn't have the enzyme glucose 6-phosphatase.

$$\text{Glucose 6-phosphate} + H_2O \underset{\text{6-phosphatase}}{\overset{\text{Glucose}}{\rightleftharpoons}} \text{Glucose} + P_i$$

In muscle, all the glucose 6-phosphate arising from glycogen degradation (via glucose 1-phosphate) either goes down glycolysis or enters the HMP pathway.

Glycogen is basically a short-term supply. In the absence of food intake, glycogen stores are depleted in about 24 hours. Over longer periods of fasting or starvation, glucose equivalents cannot be provided by glycogen stores and must come from protein sources.

Low insulin and high glucagon levels indicate that blood glucose levels are low. This stimulates glycogenolysis and inhibits glycogen synthesis. The net result is the breakdown of glycogen to glucose 1-phosphate. On the other hand, high insulin and low glucagon levels indicate that glucose levels are high and that the extra glucose should be stored as glycogen. This signal turns on glycogen synthesis and turns off glycogen degradation.

Glycogen can also help supply glucose to meet short-term demands

LONG FORM

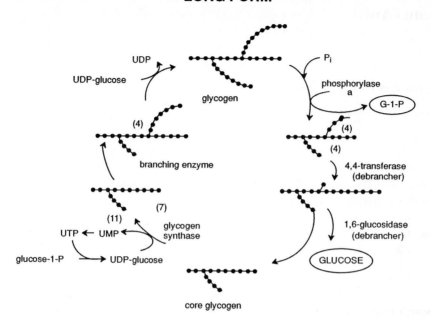

SHORT FORM

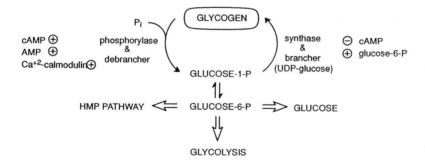

Figure 15-3   Glycogen Metabolism

for energy. The glycogen in muscle is intended to provide a short-term energy supply that can be turned on immediately in times of excitement. In muscle, epinephrine turns on glycogen degradation to provide glucose 6-phosphate for local use through glycolysis, particularly if the muscle exerts itself so much that it restricts its blood supply and becomes anaerobic. Epinephrine has the same effect on liver glycogen (↑ degradation); however, liver ships the glucose out for consumption by other tissues, including muscle.

FAT

Fat provides a long-term storage form for energy (ATP). Fat provides zero glucose equivalents.

Liver: Liver makes fat for export to other tissues.
Muscle: Resting muscle prefers fatty acids as an energy supply. However, glucose provides short-term, high-intensity supply.
Adipose: Adipose tissue is the primary storage facility of fat. Fat is stored in these tissues as an intracellular droplet of insoluble triglyceride. A hormone-sensitive lipase mobilizes triglyceride stores by hydrolysis to free fatty acids.
Red cells: These cells can't use fat at all for energy since they have no mitochondria.
Brain: Brain does not burn fat as an energy source; however, after adapting to long-term starvation, brain can use ketone bodies for fuel.

METABOLIC MOVEMENTS OF *FAT*

Feeding (↓ glucagon ↑ *insulin*):   *Fat synthesis*
Fasting (↑ *glucagon* ↓ insulin):   *Fat degradation*
Starvation (↑ glucagon ↓ insulin):   *Fat degradation*
Diabetes (↑ *glucagon* ↓ insulin):   *Fat degradation*
Excitement (↑ *epinephrine*):   *Fat degradation*

Fat is only an energy storage form (Fig. 15-4). Fat cannot be converted to carbohydrate equivalents. This is a very important point. Remember it! The reason for this is a bit subtle. The carbon skeleton of fatty acids is metabolized only to acetyl-CoA. Glucose precursors such as oxaloacetate can be synthesized from acetyl-CoA by going around the

LONG FORM

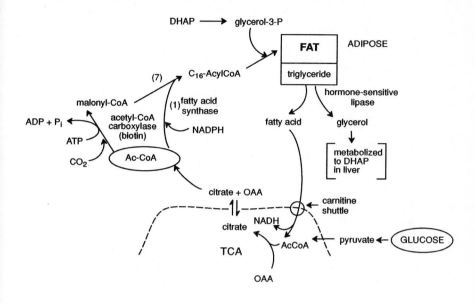

SHORT FORM

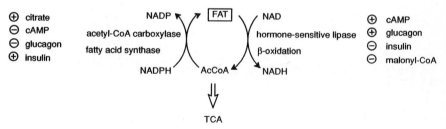

Figure 15-4   Fat Metabolism

TCA cycle. However, acetyl-CoA has 2 carbon atoms. Going around the
TCA cycle burns off 2 carbon atoms (as CO_2). The net number of carbon
atoms that ends up in oxaloacetate is then zero. No carbohydrate can be
made from fat.[4]

---

[4] Odd-chain fatty acids are an exception. While they are relatively rare in the diet, odd-
chain-length fatty acids end up at propionyl-CoA (C_3). Propionyl-CoA is carboxylated by
propionyl-CoA carboxylase to give methylmalonyl-CoA. Methylmalonyl-CoA is rearranged
to succinyl-CoA by the enzyme methylmalonyl-CoA mutase, a vitamin-B_{12}-requiring enzyme.

Fat is synthesized as long-chain fatty acids and then stored by making triglyceride. Most of the body's fat is stored in adipose tissue. While glycogen storage is limited to about a 24-hour supply, fat can be stored in unlimited amounts. Most calories in excess of those required to maintain energy demands are stored as fat. In contrast to glucose metabolism, the metabolism of fat can produce ATP only when coupled to mitochondrial respiration. Fat is an aerobic source of energy.

The formation of a triglyceride that requires the presence of glycerol introduces a carbohydrate requirement for the storage of fat. When energy and glucose equivalents are available, adipose tissue stores fat. Strangely enough, adipose tissue can synthesize the glycerol (actually the 3-phosphoglycerate) required for triglyceride synthesis. However, adipose tissue cannot simply reuse the glycerol produced from triglyceride hydrolysis. This glycerol is shipped to the liver, where it can be converted to glyceraldehyde 3-phosphate and either burned as fuel (during starvation) or used to make more triglyceride.

Many tissues (muscle, liver, renal cortex) prefer fat for an energy supply, at least in the resting state. The exception is red cells and brain. These tissues depend heavily on glycolysis for energy. Red cells cannot survive without glucose (no mitochondria), but during prolonged starvation, brain can adapt to utilize fat metabolites produced by the liver (ketone bodies).

The regulation of fat metabolism is relatively simple. During fasting, the rising glucagon levels inactivate fatty acid synthesis at the level of acetyl-CoA carboxylase and induce the lipolysis of triglycerides in the adipose tissue by stimulation of a hormone-sensitive lipase. This hormone-sensitive lipase is activated by glucagon and epinephrine (via a cAMP mechanism). This releases fatty acids into the blood. These are transported to the various tissues, where they are used.

PROTEIN

Protein is a long-term reserve of glucose and energy. Protein is also an important structural component of cells. Prolonged protein use for energy and glucose supplies depletes muscle mass.

Liver: Minor protein stores; helps muscle metabolize protein
Muscle: Major site of protein stores for metabolic needs
Adipose: No significant protein stores
Red cells: No significant protein stores
Brain: No significant protein stores

METABOLIC MOVEMENTS OF PROTEIN

Feeding (↓ glucagon ↑ *insulin*):   *Protein synthesis*
Fasting (↑ *glucagon* ↓ insulin):   *Protein degradation*
Starvation (↑ *glucagon* ↓ insulin):   *Extensive protein degradation*
Diabetes (↑ *glucagon* ↓ insulin):   *Protein degradation*
Excitement (↑ *epinephrine*):   *Protein degradation*

Protein is not just an energy-glucose storage pool (Fig. 15-5). Proteins are essential structural and functional components of cells. Rampant, nonspecific protein degradation does not occur. The process is selective. Some proteins are degraded rapidly even under normal conditions. This type of synthesis-degradation is used as a mechanism to control the metabolic pathways that use the proteins as enzymes. This cyclic synthesis-degradation does not provide a significant amount of amino acids for metabolic demands. Protein is degraded to provide for metabolic needs only as a last resort.

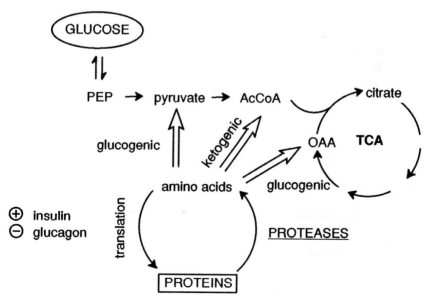

Figure 15-5   Protein Metabolism

The overriding consideration here is generating a supply of glucose equivalents for the rest of the body. Glucose equivalents are essential. When there is limited intake of glucose (fasting) or limited capability to utilize glucose (diabetes), protein is degraded to provide this essential supply of glucose. Each amino acid has its own pathway for degradation. Some amino acids are degraded either to pyruvate or to one of the intermediates of the TCA cycle. These are called the *glucogenic* amino acids, and it's these that we're after if we don't have glucose available. Unfortunately, we have to degrade the whole protein to get at them.

The glucogenic amino acids are those that produce degradation products that can be converted *only to* glucose (there are a bunch of these). Two of the amino acids (Leu and Lys) are degraded only to acetyl-CoA or molecules that cannot be used to synthesize glucose. These are the *ketogenic* amino acids. Some amino acids are degraded to different products, one of which is ketogenic and one of which is glucogenic. These amino acids (Ile, Phe, Tyr, Trp) are *both ketogenic and glucogenic.*

Protein may also be degraded in order to supply essential amino acids for the synthesis of more essential proteins. This may occur in the case of an essential–amino acid deficiency. If the dietary intake of one or more essential amino acids is insufficient, nonessential proteins are degraded to provide this supply. This may result not only in protein degradation but also in the oversupply of nonessential amino acids and their metabolites. The degradation of proteins to make up for a limited supply of an essential amino acid causes a negative nitrogen balance. Nitrogen balance is essentially nitrogen in minus nitrogen out. When protein is degraded because of a short supply of an essential amino acid, the extra amino acids that you get in the process must be degraded, and the nitrogen has to go somewhere—usually out as urea and ammonia. More nitrogen out than nitrogen in means that protein is being degraded. This phenomenon can be used to tell if an amino acid is essential or not: Delete it from the diet and watch for negative nitrogen balance.

The primary location for metabolically useful protein storage is muscle. As long as sufficient glucose is available, either from the diet or from glycogen degradation, protein is spared from degradation. Significant protein degradation for metabolic needs occurs primarily upon long-term starvation or in cases such as uncontrolled diabetes, which is effectively the same as starvation. Protein is constantly being degraded; however, it is constantly being synthesized. Normally, the processes are balanced and the amount of protein stays relatively constant. Although protein in other tissues is degraded under conditions of starvation, muscle protein provides the most mass and the most extensive protein pool for metabolic needs.

TISSUE COOPERATION

Individual tissues have specific roles in maintaining a constant supply of glucose and energy for the whole organism.

LIVER

The liver maintains blood glucose levels, makes fat, makes ketone bodies, and stores glycogen.

Feeding: Stores glycogen and makes fat
Fasting: Degrades glycogen, makes glucose, and burns fat
Starving: Makes glucose, burns fat, makes ketone bodies, burns amino acids, and handles nitrogen
Diabetes: Makes glucose, burns fat, makes ketone bodies, burns amino acids, and handles nitrogen
Excitement: Degrades glycogen and makes glucose

MUSCLE

Muscle provides movement, stores glycogen for its own use, and stores protein.

Feeding: Stores glycogen, burns fat and glucose, and makes protein
Fasting: Degrades glycogen and burns fat
Starving: Burns fat and ketone bodies and degrades proteins
Diabetes: Burns fat and ketone bodies and degrades proteins
Excitement: Degrades glycogen and burns glucose

ADIPOSE

Adipose tissue manages fat stores.

Feeding: Stores fat
Fasting: Releases stored fat
Starving: Releases stored fat
Diabetes: Releases stored fat
Excitement: Releases stored fat

BRAIN

The brain thinks (hopefully).

Feeding: Uses glucose
Fasting: Uses glucose
Starving: Induces enzymes to use ketone bodies
Diabetes: Induces enzymes to use ketone bodies
Excitement: Sits back and enjoys it

CONNECTION OF STORAGE POOLS

The flow of energy through glycogen, fat, and protein makes sense. For the most part, the players you'll see are the hormones insulin, glucagon, and epinephrine and the interrelationships that they orchestrate between the various tissues. The reason for all this complexity is, for the most part, because we don't eat constantly. We have food coming in several times a day, and this has to last until we eat again. Immediate dietary supplies are exhausted in a few hours after eating. Glycogen stores are depleted about 24 hours after the last meal. Assuming an adequate water supply, individuals have survived starvation conditions for long periods. The various metabolic pathways and tissue-cooperative effects are set up to keep us fed between meals and to protect us as long as possible in the absence of food.

The connections between the storage pools and the directions of flow depend on the metabolic condition. There are a couple of generalities.

Cells always need energy. There will always need to be some utilization of fat or glucose for energy through the TCA cycle. Metabolic switches (control points) are not usually on–off switches—they are up–down switches. Flux through a pathway is rarely zero, even when the pathway is "off." The purpose of the whole scheme is to maintain a constant supply of energy and glucose. When glucose and energy are plentiful, they are stored, and the flux of metabolism is increased in the direction of storage. When glucose and energy are needed and cannot be obtained directly from dietary intake, the carbon flux shifts to withdraw material from storage. Other pathways, such as the Cori and alanine cycles, ketone body synthesis and utilization, and increased nitrogen metabolism, are adaptations that use cooperation between various tissues and organs to enhance the flow of energy and glucose.

FEEDING

Signals:
  Insulin: *Up*
  Glucagon: *Down*
Storage forms:
  Glycogen: *Up*
  Fat: *Up*
  Protein: *Up*

Right after feeding, the digestive system is breaking down food and dumping nutrients (amino acids, free fatty acids, and carbohydrate) into the portal system and then to liver and ultimately into the rest of the circulation (Fig. 15-6). The key here is that all tissues take in these nutrients, burn only what they need at the moment, and store the rest.

One of the major signals here is an increase in the level of insulin. This tells the tissues that glucose is abundant, and it increases the flow of glucose into cells. At the same time, the glucagon level is falling. This is also a signal for an adequate supply of glucose in the blood. Because of this, cAMP levels are low and regulated proteins become less phosphorylated.

All tissues are doing essentially the same thing: taking in glucose, fat, and amino acids and storing them. All tissues have some storage capacity. However, for the most part, different organs and tissues store different things. Liver stores glycogen, adipose tissue stores fat, and muscle stores glycogen (for itself) and protein.

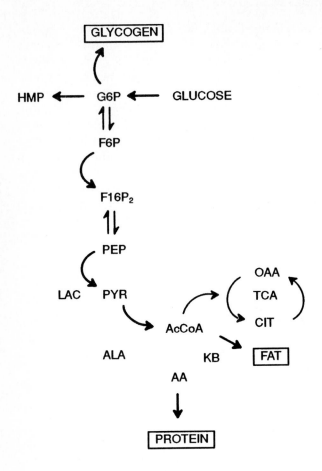

LIVER, MUSCLE, ADIPOSE

GLYCOGEN INCREASES
GLYCOLYSIS
FAT SYNTHESIS
PROTEIN SYNTHESIS

Figure 15-6   Metabolic Movements after Feeding
Synthesis and storage pathways are on; degradative pathways are off. AA =
amino acids; KB = ketone bodies; ALA = alanine; LAC = lactate.

FASTING

Signals:
  Insulin: *Down*
  Glucagon: *Up*

Storage forms:
  Glycogen: *Down*
  Fat: *Down*
  Protein: *Down*

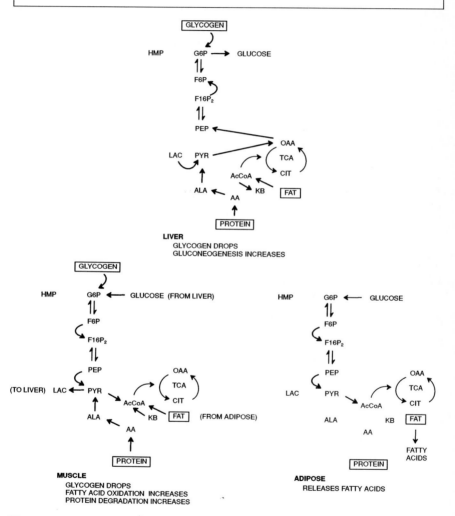

Figure 15-7 Metabolic Movements after Fasting
Storage pathways are off; degradative pathways are on. AA = amino acids;
KB = ketone bodies; ALA = alanine; LAC = lactate.

A few hours after eating, the supply of easily available nutrients has been used up and stores must be mobilized to maintain relatively constant supplies of energy and glucose (Fig. 15-7). Insulin levels drop and glucagon levels rise. Because of the increased glucagon levels, cAMP levels rise and regulated proteins become more phosphorylated.

Liver's main job is to keep up the levels of blood glucose. To do this, it breaks down glycogen and turns on gluconeogenesis. The liver takes lactate and alanine from the circulation and through gluconeogenesis converts it into glucose. The ammonia from the alanine is pushed through the urea cycle. For its own energy supply, the liver takes in free fatty acids (released from the adipose tissue) and metabolizes them to acetyl-CoA. Fatty acid synthesis is inhibited. The increased reliance on fat for energy and the increased supply of acetyl-CoA increases ketone body synthesis somewhat, but not to the extent that is observed during starvation.

Muscle breaks down glycogen during fasting to meet its own needs. Skeletal muscle is missing the enzyme glucose 6-phosphatase, and it can't generate any free glucose for the rest of the body to use. The lactate from glycolysis is shipped to the liver. Muscle also picks up free fatty acids and metabolizes them to acetyl-CoA and on through the TCA to CO_2.

Adipose tissue releases fat by activation of the hormone-sensitive lipase. The glycerol released by adipose tissue can provide some glucose equivalents to the liver. Adipose tissue itself can't use glycerol—it's missing glycerol kinase to make glycerol 3-phosphate.

STARVATION

Signals: **Storage forms:**
  Insulin: *Down*   Glycogen: *Exhausted*
  Glucagon: *Up*   Fat: *Down*
                                        Protein: *Down*

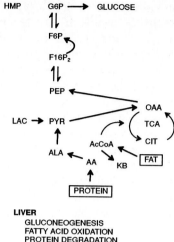

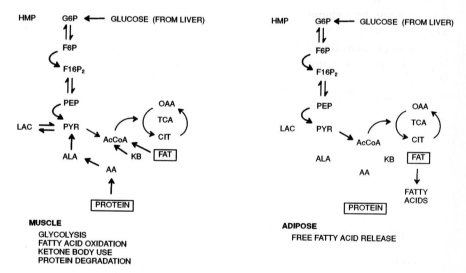

Figure 15-8   Metabolic Movements during Starvation
Storage pathways are off; degradative pathways are on. Glycogen has been depleted, and maintaining glucose levels becomes a big problem. AA = amino acids; KB = ketone bodies; ALA = alanine; LAC = lactate.

Glycogen stores in liver and kidney are exhausted in about 24 hours. After this, the body must find glucose equivalents somewhere. The major metabolic adaptations of starvation are the result of having to maintain glucose levels without any direct source of it (Fig. 15-8).

During starvation, liver continues in the gluconeogenic mode. It's making glucose for the rest of the body. Lactate and alanine from other tissues are taken in and used through gluconeogenesis to make glucose. Protein is degraded to get at amino acids that can be degraded into glucose equivalents. Fat is increasingly relied on for energy. This excess demand for β oxidation results in the increased synthesis of ketone bodies, which are made by the liver and exported to the other tissues.

Because it lacks glucose 6-phosphatase activity, muscle can't directly participate in maintaining glucose levels during starvation. However, the increased availability of glucogenic amino acids from the increased degradation of muscle protein can be used by the liver to synthesize glucose. Muscle also utilizes ketone bodies (and fat) for energy.

Adipose tissue continues to dump free fatty acids into the circulation until its supplies are exhausted.

Brain, which is usually very reliant on glucose for energy, adapts in a few days of starvation to use ketone bodies as a source of energy. This spares the body some glucose, which is still essential to maintain red cell function.

EXCITEMENT

Signals:
  Epinephrine: *Up*
Storage forms:
  Glycogen: *Down*
  Fat: *Down*
  Protein: *Down*

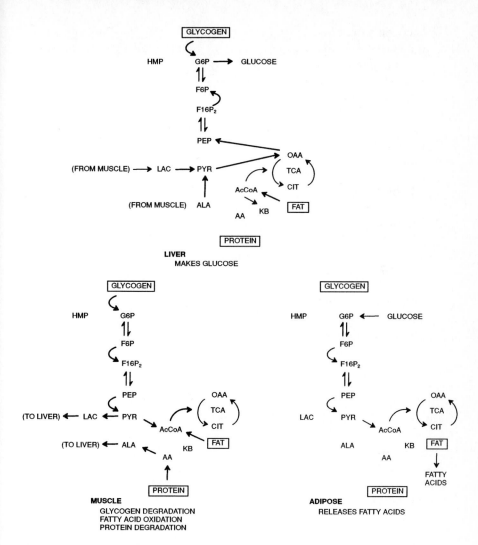

Figure 15-9   Metabolic Movements during Excitement
Storage pathways are off; degradative pathways are on. Glycogen is degraded by
liver and muscle to provide glucose and energy. AA = amino acids; KB = ketone
bodies; ALA = alanine; LAC = lactate.

Excitement produces a short-term demand for energy and glucose by the muscle (Fig. 15-9). In the muscle, epinephrine stimulates glycogenolysis and glycolysis to provide immediate fuel and ATP. But the other tissues help out. In the liver, epinephrine stimulates glycogenolysis and gluconeogenesis in order to maintain an adequate supply of glucose to meet the increased demands of muscle. Fatty acid oxidation is stimulated to supply energy. Adipose tissue is stimulated to release free fatty acids for the other tissues.

Note that the effects of epinephrine on glycolysis are different in the liver and muscle. This is because liver must make glucose under these conditions, not burn it. The different responses of liver and muscle come from a difference in the effect of phosphorylation on the enzyme that makes and degrades fructose 2,6-bisphosphate, a major regulator of glycolysis and gluconeogenesis. In muscle, glycolysis is stimulated by epinephrine. Since epinephrine causes increased phosphorylation of proteins and since the concentration of fructose 2,6-bisphosphate must rise to activate glycolysis, phosphorylation in muscle must activate the enzyme that makes fructose 2,6-bisphosphate and inactivate the enzyme that hydrolyzes it. In liver, gluconeogenesis is activated by epinephrine. Since epinephrine causes increased phosphorylation of proteins and since the concentration of fructose 2,6-bisphosphate must fall to activate gluconeogenesis, phosphorylation in the liver must activate the enzyme that hydrolyzes fructose 2,6-bisphosphate and inhibit the enzyme that makes it. Believe it or not, this is actually what happens. The liver and muscle forms of the kinase that makes fructose 2,6-bisphosphate show exactly the opposite effects of phosphorylation on activity. Note that in order for you to be able to reconstruct all the effects of phosphorylation of all these enzyme activities all you needed to know was two facts: cAMP increases protein phosphorylation and fructose 2,6-bisphosphate activates glycolysis and inhibits gluconeogenesis.

INTERORGAN CYCLES

Several interogan cycles of carbon flow shift the metabolic burden from one tissue type to another.

CORI CYCLE

Muscle uses glucose from the liver to make lactate.
Liver uses lactate from muscle to make glucose.

In the Cori cycle (Fig. 15-10), the muscle produces lactate from glucose as a result of using glycolysis for energy. If glycolysis is faster than the ability of the muscle to metabolize the pyruvate aerobically through the TCA cycle, the pyruvate is reduced to lactate and exported from the muscle. The lactate circulates through the blood to the liver, where gluconeogenesis converts the lactate to glucose. The glucose travels through the blood back to the muscle, where it is reutilized. The recycling of the lactate extends the supply of glucose stores and provides an indirect way to get muscle glycogen into the glucose pool. The cycle is not without cost. It's not a perpetual motion machine.

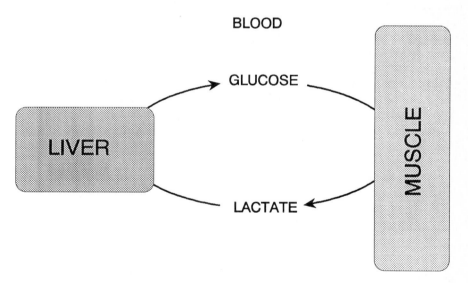

Figure 15-10   The Cori Cycle
Cooperation between liver and muscle recycles lactate into glucose.

The conversion of glucose to lactate produces 2 ATPs for muscle:

$$\text{Glucose} + 2\text{ADP} + 2\text{P}_i \longrightarrow 2 \text{ lactate} + 2\text{ATP}$$

However, the conversion of lactate to glucose requires the expenditure of 6 ATP equivalents by the liver:

2 lactate + 2~~NAD~~$^+$ $\longrightarrow$ ~~2 pyruvate~~ + ~~2NADH~~
~~2 pyruvate~~ + 2ATP + ~~2CO$_2$~~ $\longrightarrow$ ~~2 oxaloacetate~~ + 2ADP + 2P$_i$
~~2 oxaloacetate~~ + 2GTP $\longrightarrow$ ~~2PEP~~ + 2GDP + 2P$_i$ + ~~2CO$_2$~~
~~2PEP~~ + 2ATP + ~~2NADH~~ $\longrightarrow$ glucose + 2ADP + 2P$_i$ + ~~2NAD~~$^+$
_____

Net: 2 lactate + 4ATP + 2GTP $\longrightarrow$ 4ADP + 2GDP + 6P$_i$

Each mole of glucose that goes through the Cori cycle costs the liver 6 ATP equivalents. If the Cori cycle were perfect and there were an endless supply of ATP, liver could supply glucose equivalents forever just using the same carbon atoms (6 in as lactate, 6 out as glucose). However, some of the lactate (pyruvate) in the nonliver tissues is burned to CO_2 by the TCA cycle in the muscle and other tissues. It's this pool that must be replaced by fresh glucose from liver.

ALANINE CYCLE

Liver takes carbon and nitrogen waste from muscle (alanine), disposes of the nitrogen, and recycles the carbon into glucose.

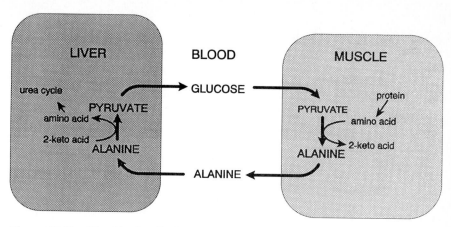

Figure 15-11   The Alanine Cycle
Cooperation between liver and muscle allows muscle to get rid of nitrogen waste
and recycle the carbon skeleton into glucose.

The alanine cycle accomplishes the same thing as the Cori cycle,
except with an add-on feature (Fig. 15-11). Under conditions under which
muscle is degrading protein (fasting, starvation, exhaustion), muscle must
get rid of excess carbon waste (lactate and pyruvate) but also nitrogen
waste from the metabolism of amino acids. Muscle (and other tissues)
removes amino groups from amino acids by transamination with a 2-keto
acid such as pyruvate (oxaloacetate is the other common 2-keto acid).

$$CH_3C(\!=\!O)CO_2^- \ + \ ^-O_2C\!-\!CH(NH_3^+)CH_2CH_2\!-\!CO_2^+$$

$$\text{Pyruvate} \qquad\qquad\qquad \text{Glutamate}$$

$$CH_3\!-\!CH(NH_3^+)\!-\!CO_2^- \ + \ ^-O_2C\!-\!C(\!=\!O)CH_2CH_2CO_2^-$$

$$\text{Alanine} \qquad\qquad\qquad \alpha\text{-Ketoglutarate}$$

The result is that the amino groups can be dumped out as alanine (the
transamination product of pyruvate). In the liver and kidney, alanine is
transaminated to yield pyruvate and glutamate. As in the Cori cycle, the
pyruvate is converted to glucose by the liver and is shipped out. The
glutamate is fed into the urea cycle–nitrogen disposal system to get rid of
the excess nitrogen.

KETONE BODIES

Acetoacetate: $CH_3C(=O)CH_2CO_2^-$
β-Hydroxybutyrate: $CH_3CH(OH)CH_2CO_2^-$
Formed in liver to free up CoA for β oxidation
Metabolized in other tissues, including brain, as an energy source

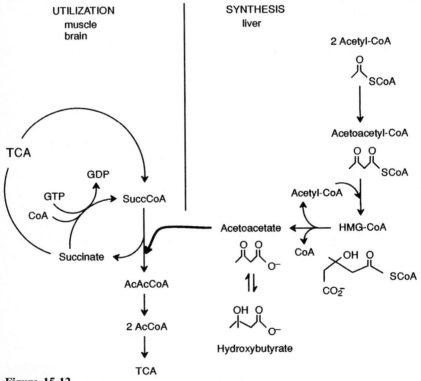

Figure 15-12
KETONE BODIES are generated by the liver and used by muscle and brain (after adaptation during starvation).

The formation of ketone bodies is a consequence of prolonged metabolism of fat (Fig. 15-12). Their formation in the liver actually enables liver to metabolize even more fat by freeing up CoA that would otherwise be tied up as acetyl-CoA waiting to get into the TCA cycle. The liver exports the ketone bodies, and other tissues, particularly brain, can adapt to use them.

With increasing metabolism of fat through β oxidation, much of the mitochondrial CoA pool may become tied up as acyl- or acetyl-CoA. In such cases, the supply of free CoA can be diminished, and this may limit the rate of β oxidation. Upon prolonged fasting and heavy reliance on fat for energy, the liver induces the enzymes required for the formation of ketone bodies and brain induces enzymes required for their metabolism.

Ketone bodies are formed in the liver mitochondria by the condensation of three acetyl-CoA units. The mechanism of ketone body formation is one of those pathways that doesn't look like a very good way to do things. Two acetyl-CoAs are condensed to form acetoacetyl-CoA. We could have had an enzyme that just hydrolyzed the acetoacetyl-CoA directly to acetoacetate, but no, it's got to be done in a more complicated fashion. The acetoacetyl-CoA is condensed with another acetyl-CoA to give hydroxymethylglutaryl-CoA (HMG-CoA). This is then split by HMG-CoA lyase to acetyl-CoA and acetoacetate. The hydroxybutyrate arises from acetoacetate by reduction. The overall sum of ketone body formation is the generation of acetoacetate (or hydroxybutyrate) and the freeing-up of the 2 CoAs that were trapped as acetyl-CoA.

Ketone bodies are utilized in other tissues (not liver) by converting the acetoacetate to acetoacetyl-CoA and then converting the acetoacetyl-CoA to 2 acetyl-CoA, which are burned in the muscle mitochondria.

UREA CYCLE

·

· · · · · · · · · · · · ·

UREA CYCLE

Function: To provide a route to dispose of the amino groups from amino acids during their metabolism

Location: Liver

Connections: *From* amino groups of amino acids through glutamate and glutamate dehydrogenase

From amino groups of amino acids through aspartate and ar-gininosuccinate synthase

From ammonia through carbamoyl phosphate synthetase

To urea

Regulation: Primarily by availability of amino groups and ammonia

Equation:

$$NH_4^+ + CO_2 + Asp + 2ATP \longrightarrow$$
$$NH_2C(\!\!=\!\!O)NH_2 + fumarate + 2ADP + 2P_i$$

$$Glu + NAD^+ + CO_2 + Asp + 2ATP \longrightarrow$$
$$NH_2C(\!\!=\!\!O)NH_2 + \alpha\text{-ketoglutarate} + fumarate + 2ADP + 2P_i$$

(See Fig. 16-1.)

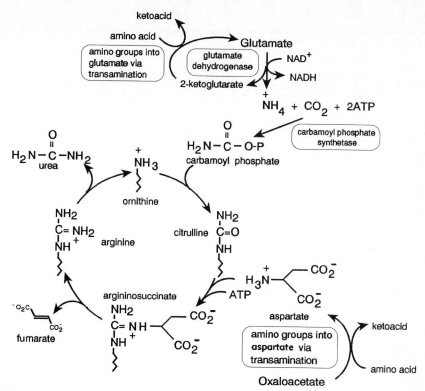

Figure 16-1   The Urea Cycle

PURINE METABOLISM

·

Purine Synthesis

Purine Salvage

Deoxynucleotides

Purine Degradation

PURINE SYNTHESIS

Function: To provide purines (A and G) for energy metabolism and for DNA-RNA synthesis.

Location: Everywhere.

Connections: To folate metabolism and one-carbon metabolism in de novo synthesis.

From HMP pathway via ribose and PRPP.

To deoxyribonucleotides through ribonucleotide reductase.

Regulation: Availability of PRPP.

Activity of the enzyme catalyzing the formation of the 5-phosphoribosyl-1-amine from PRPP is inhibited by purines.

Synthesis of GMP requires ATP.

Synthesis of AMP requires GTP.

Equation:

PRPP + glutamine + glycine + formyl-THF + aspartate
$$+ \text{ some ATP} \longrightarrow \text{purines}$$

(See Fig. 17-1.)

PURINE SALVAGE

HGPRTase: Hypoxanthine or guanine + PRPP $\longrightarrow$
$$IMP \text{ or } GMP + PP_i$$
APRTase: Adenine + PRPP $\longrightarrow$ AMP + PP_i

The free bases of the purines can be salvaged to spare de novo synthesis. The only hard thing is remembering what the names stand for. HGPRTase is hypoxanthine-guanine phosphoribosyltransferase, and it makes both IMP and GMP. A separate enzyme exists for the salvage of adenine. The salvage pathways are included in Fig. 17-1.

DEOXYNUCLEOTIDES

NDP + thioredoxin(SH)$_2$ $\underrightarrow{\text{ribonucleotide reductase}}$
$$dNDP + \text{thioredoxin(SS)}$$
dTMP made from dUMP with CH_2—THF

Deoxynucleotides for DNA synthesis are made at the nucleoside *diphosphate* level and then have to be phosphorylated up to the triphosphate using a kinase and ATP. The reducing equivalents for the reaction come from a small protein, thioredoxin, that contains an active site with two cysteine residues. Upon reduction of the ribose to the 2'-deoxyribose, the thioredoxin is oxidized to the disulfide. The thioredoxin(SS) made during the reaction is recycled by reduction with NADPH by the enzyme thioredoxin reductase.

Ribonucleotide reductase works on ribo-A, -U, -G, -C diphosphates to give the deoxynucleotide. The deoxyuridine, which is useless for RNA synthesis, is converted to deoxythymidine by the enzyme thymidylate synthase, which uses methylene tetrahydrofolate as a one-carbon donor. The odd thing here is that ribonucleotide reductase uses the UDP as a substrate to give the dUDP. This must then by hydrolyzed to the dUMP before thymidylate synthase will use it to make dTMP. Then the dTMP has to be kinased (phosphorylated) up to dTTP before DNA can be made.

Regulation of ribonucleotide reductase is a bear. There appear to be two regulatory sites, one that affects the overall activity and another that changes the relative specificity for the various purine and pyrimidine substrates. ATP binding to the activity regulatory site turns on the ac-

PURINES

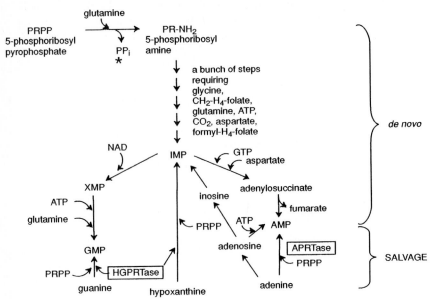

Figure 17-1   Purine Synthesis and Salvage

tivity toward all substrates, while dATP binding to the same site turns it off. A complex pattern of specificity changes is observed when one of the deoxynucleotides binds to the second regulatory site. The general idea is to keep the levels of the various deoxynucleotides at the proper levels and ratios for DNA synthesis. ATP binding to the specificity site activates the formation of dCDP, while dTTP binding to the site activates dGDP formation and inhibits dCDP formation. The dGTP activates the formation of dADP but inhibits the formation of dCDP and dGDP. Simple . . . ? Don't put this one too high on your trivia list.

PURINE DEGRADATION

GMP → guanosine → guanine → xanthine → urate
AMP → adenosine → inosine → hypoxanthine → xanthine → urate
AMP → IMP → inosine → → → urate

Name changes may be confusing here; when AMP loses the phosphate to become adenosine and adenosine loses the ribose to become adenine, it's still easy to tell who came from where. When IMP loses the phosphate, it becomes inosine, but when inosine loses the ribose it becomes hypoxanthine. It may be a little confusing, but it's still better than trying to pronounce *inonine*.

PYRIMIDINE
METABOLISM

·

Pyrimidine Synthesis

Pyrimidine Salvage

Pyrimidine Degradation

· · · · · · · · · · ·

PYRIMIDINE SYNTHESIS

Function: To make pyrimidine nucleotides (U, T, C) for DNA and RNA synthesis.

Location: Everywhere.

Connections: To amino acid metabolism by the requirement for glutamine and aspartate.

Regulation: UTP inhibits synthesis of carbamoylphosphate.

Equation:

$$Gln + CO_2 + Asp + PRPP + some\ ATP \longrightarrow UTP + CTP$$

The major difference between purine and pyrimidine de novo biosynthesis is that the pyrimidine ring is assembled and then added to PRPP (Fig. 18-1). With purines, the purine ring is built directly on the PRPP.

PYRIMIDINES

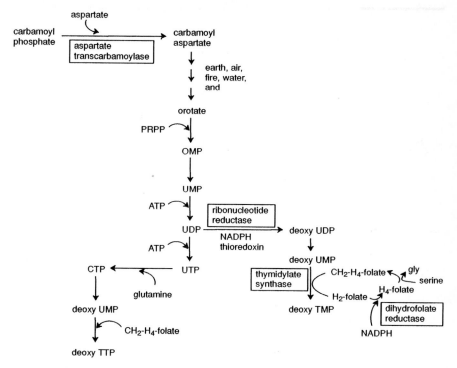

Figure 18-1   Pyrimidine Synthesis and Salvage

PYRIMIDINE SALVAGE

Uracil + PRPP $\longrightarrow$ UMP + PP$_i$ (only uracil)

Nucleoside phosphorylase:

U, C, T + ribose 1-phosphate $\longrightarrow$ nucleosides + P$_i$

There are basically two types of salvage. The first involves attachment of the base to PRPP with the formation of pyrophosphate. This pathway is available for salvage of purines and uracil but not for cytosine or thymine. The other pathway involves attachment of the base to ribose 1-phosphate, which occurs to some extent for most of the purines and pyrimidines. This second pathway requires the presence of specific kinases to convert the nucleoside to the monophosphate. Except for thymidine kinase, these kinases do not exist in most cells.

PYRIMIDINE DEGRADATION

CMP $\longrightarrow$ cytidine $\longrightarrow$ uridine $\longrightarrow$ uracil $\longrightarrow$ $\longrightarrow$
$\quad\quad\quad\quad\quad$ β-alanine $\longrightarrow$ malonate semialdehyde $\longrightarrow CO_2$

Thymidine $\longrightarrow$ thymine $\longrightarrow$ β-aminobutyrate $\longrightarrow$
$\quad\quad\quad\quad\quad\quad$ methylmalonate semialdehyde $\longrightarrow CO_2$

After removal of the phosphates by various phosphatases, the nucleosides are cleaved to the base by the same nucleoside phosphorylase that catalyzes the salvage reaction. The equilibrium constant for this reaction is near 1, so that it can go in either direction depending on the relative levels of the substrates and products.

$$\text{Base–ribose} + P_i \rightleftharpoons \text{base} + \text{ribose 1-phosphate}$$

The nitrogen from the pyrimidine bases is removed by transamination and dumped onto glutamate. The carbon skeleton ends up as CO_2.

ONE-CARBON METABOLISM

·

One-Carbon Metabolism

Oxidation States of Carbon

· · · · · · · · · · · · ·

ONE-CARBON METABOLISM

Function: To donate methyl groups to phospholipid, biogenic amines, thymidine, and amino acid biosynthesis
To provide one-carbon fragments at the level of formaldehyde and formic acid for purine and pyrimidine biosynthesis
Location: Most everywhere
Connections: One-carbon fragments *in* from serine, glycine, formate, and histidine
One-carbon fragments *out* from SAM, formyl-THF, methylene-THF, and methyl-THF
Regulation: At individual enzyme level

(See Fig. 19-1.)

OXIDATION STATES OF CARBON

Count the number of carbons and hydrogens connected to the carbon in question. Carbon–carbon double bonds count only once. The lower the number, the more oxidized the carbon. Conversions between levels require oxidizing or reducing agents. Conversions within a given level require no oxidizing or reducing agents.

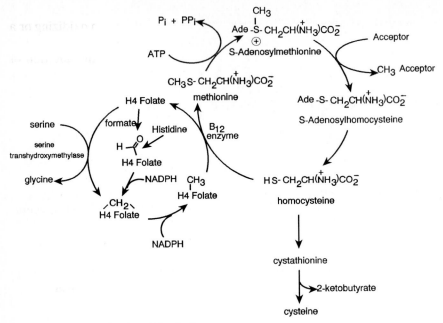

Figure 19-1   One-Carbon Metabolism

To determine the oxidation state of a specific carbon atom is simple. Just count the number of carbon and hydrogen atoms that the carbon atom in question is connected to. Carbon–carbon double bonds count only once. A more reduced carbon has a higher number, and a more oxidized carbon has a lower number. Carbon atoms can be in five different oxidation states.[1] Being in a different oxidation state means that some source of oxidizing or reducing agent must be used to convert carbon in one oxidation state to carbon in another oxidation state. In terms of the table above, this means that to move up in the table (to a more reduced form of carbon) requires a reducing agent like NADH. Moving down the table requires an oxidizing agent like NAD^+ or oxygen. Moving between successive oxidation states represents a two-electron oxidation or reduction. Conversion of carbon within a given redox state does not require an oxidizing or reducing agent.

For example, converting methylene-THF [$-N-CH_2-N-$, state 2][2] to formyl-THF [$-N-C(=O)-H$, state 1] would require an oxidizing agent.

---

[1] This doesn't count carbon atoms with single electrons (free radicals). You've got to draw the line somewhere, and I've chosen to eliminate the more radical elements. If you want to put them in, you can draw your own table.

On the other hand, conversion of formyl-THF [$-N-C(=O)-H$, state 1][2] to methenyl-THF ($-N-CH=N-$, state 1) would not require an oxidizing or a reducing agent. The way to think about the conversion between methenyl-THF and formyl-THF is that the reaction is simply the addition of another amino group from the THF to the $C=O$ of the formyl group followed by the elimination of water. In none of the reactions does the carbon atom change its oxidation state.[2]

$$-N-CH(=O) + NH_2-R \rightleftharpoons -N-CH(OH)-NH-R$$

$$-N-CH(OH)-NH-R \rightleftharpoons -N-CH=N-R + H_2O$$

On the other hand the conversion of methenyl-THF ($-N-CH=N-R$, state 1) to methylene-THF ($-N-CH_2-N-$, state 2) requires a reducing agent, NADPH.

Reduction Level	Name	Typical Structures	Folic Acid Equivalent[2]
4	Methane	CH_4 CH_3-C $C-CH_2-C$	None
3	Methanol	CH_3OH CH_3Cl $CH_2=C-$	Methyl-THF ($-N-CH_3$)
2	Formaldehyde	$H-C(=O)-H$ $H-C(OH)_2-H$	Methylene-THF ($-N-CH_2-N-$)
1	Formic acid	$H-C(=O)-OH$	Formyl-THF ($-N-C(=O)-H$) Methenyl-THF ($-N-CH=N-$)
0	Carbon dioxide	$O=C=O$ $HO-C(=O)-OH$ $H_2N-C(=O)-NH_2$	None

[2] The structural features shown in parentheses or brackets represent the structure of the one-carbon fragment attached to the N^5 and N^{10} of tetrahydrofolate. The bonds to carbon are as shown, but for simplicity all the bonds to N may not be shown.

TRACKING CARBONS

·

Glucose to Pyruvate

TCA cycle

· · · · · · · · · · · · ·

GLUCOSE TO PYRUVATE

Glucose:

(O=)CH–CH(OH)–CH(OH)– | –CH(OH)–CH(OH)–CH$_2$(OH)

$\quad$ *1* $\quad$ 2 $\qquad$ 3 $\qquad\qquad$ 4 $\qquad$ 5 $\qquad$ 6

$\qquad\qquad\qquad$ *1,6* $\quad$ *2,5* $\quad$ *3,4* $\qquad\qquad$ Glucose numbers

Pyruvate $\qquad$ CH$_3$—C(=O)—CH$_2^-$

$\qquad\qquad\qquad$ 3 $\qquad$ 2 $\qquad$ 1 $\qquad\qquad$ Pyruvate numbers

$\qquad\qquad\qquad$ *1,6* $\quad$ *2,5* $\qquad\qquad$ *3,4* Glucose numbers

Acetyl-CoA $\quad$ CH$_3$—C(=O)—ScoA + CO$_2$

$\qquad\qquad\qquad$ 2 $\qquad$ 1

$\qquad\qquad\qquad\qquad\qquad\qquad\qquad$ Acetyl-CoA numbers

---

$\qquad$Tracing labeled carbon atoms through metabolic pathways would, at first glance,[1] appear to be a pretty irrelevant thing to make you do. But if you've got to do it, there are a couple of conceptual tricks that make it somewhat easier.

$\qquad$The first concept is that organic compounds are numbered starting with the end of the molecule that is closest to the most oxidized carbon. In glucose, C-1 is the carbon of the aldehyde. In fructose

---

[1] The second and third glances may appear this way too.

[CH$_2$(OH)–C(=O)–CH(OH)–CH(OH)–CH(OH)–CH$_2$(OH)], C-1 is the carbon at the end closest to the C=O; the C=O itself is C-2.

Labeling a given molecule at a specific carbon atom means that an isotopically labeled carbon atom (^{14}C or ^{13}C) has been introduced at a given position in the starting molecule. For example, the statement that glucose has been labeled at C-2 means that only carbon number 2 of glucose has been tagged with a ^{14}C or ^{13}C carbon atom. The other carbon atoms are not labeled (they are ^{12}C). The problem is to determine where this label ends up after the glucose is metabolized to some other compound.

Tracing the carbons of glucose to pyruvate gets complicated when fructose 1,6-bisphosphate (FBP) is cleaved into two 3-carbon fragments, glyceraldehyde 3-phosphate (G3P) and dihydroxyacetone phosphate (DHAP). The numbers of the original carbons of *glucose* are indicated by the superscripts next to the carbon.

^{1}CH$_2$OP
|
^{2}C=O
|
^{3}CHOH
+
^{4}CHOH
|
^{5}CHOH
|
^{6}CH$_2$OP
FBP

→ **Aldolase** →

^{1}CH$_2$OP
|
^{2}C=O
|
^{3}CH$_2$OH
DHAP
+
^{4}CH(=O)
|
^{5}CHOH
|
^{6}CH$_2$OP
G3P

→ **TIM** →

^{1}CH$_2$OP
|
^{2}CHOH
|
^{3}CH(=O)
G3P

↘ ↘

Pyruvate Number

^{3,4}CO$_2^-$ 1
|
^{2,5}C=O   2
|
^{1,6}CH$_3$ 3

The guilty party is the triose phosphate isomerase (TIM) reaction that interconverts DHAP and G3P. To be converted to pyruvate, the DHAP first has to be converted to G3P. TIM just moves the carbonyl group between the two carbons that don't have phosphate attached. TIM doesn't touch the phosphate. So, if the DHAP is labeled at the carbon that has the phosphate attached, the G3P that comes from DHAP will be labeled at the carbon with the phosphate attached. The carbon with the phosphate attached in the G3P that was produced directly by the aldolase reaction came from C-6 of glucose, but the carbon with the phosphate attached in the G3P that was produced from DHAP came from C-1 of glucose. After TIM does its stuff, the carbon of G3P that has the phosphate will be labeled if *either C-1 or C-6* of the original glucose was labeled. This all makes sense if you remember that pyruvate only has three carbons but glucose has six. Each carbon atom of pyruvate must come from two different carbons of the original glucose molecule.

The carboxylate group of pyruvate, being the most oxidized, is called C-1, and the CH_3 group is called C-3. The carboxylate group of pyruvate comes from the aldehyde of G3P, so C-4 and C-3 of glucose end up at C-1 of pyruvate. There is an easy way to remember which carbons of glucose end up on the same carbon of pyruvate—the numbers of the equivalent carbons sum to 7.

The labeling trick also works backward. You could have to decide which carbons of glucose become labeled when you use pyruvate labeled on a given carbon. Labeling pyruvate on C-1 will result in a glucose molecule that is labeled on both C-3 and C-4, again because of the TIM reaction.

TCA CYCLE

The two carbons entering from acetyl-CoA do not leave as CO_2 on the first cycle.
Carbons 2 and 3 of succinate are equivalent.
Carbons 1 and 4 of succinate are equivalent.

To actually understand how labeled carbons travel through the TCA cycle, you have to draw out the chemical structures of the cycle members, put a tiny little asterisk by the labeled carbon, and follow it around and around. There are two concepts, however, that you have to know in order to do this. First, the two carbons entering the cycle as acetyl-CoA are not lost as CO_2 during the first turn of the TCA cycle.

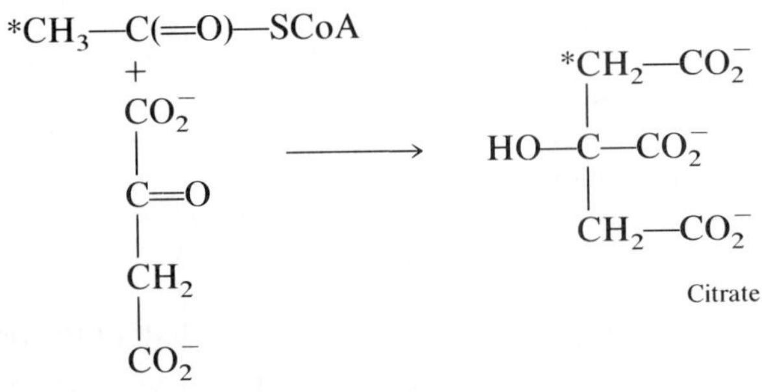

$$*CH_3 \!\!-\!\! C(\!\!=\!\!O)\!\!-\!\! SCoA$$
$$+$$
$$CO_2^-$$
$$|$$
$$C\!\!=\!\!O$$
$$|$$
$$CH_2$$
$$|$$
$$CO_2^-$$

Oxaloacetate

$\longrightarrow$

$$*CH_2\!\!-\!\!CO_2^-$$
$$|$$
$$HO\!\!-\!\!C\!\!-\!\!CO_2^-$$
$$|$$
$$CH_2\!\!-\!\!CO_2^-$$

Citrate

The two CH_2-CO_2^- arms of citrate are different (Fig. 20-1). They can be (and are) distinguished by the aconitase enzyme. To see that they're different, imagine grabbing the CO_2^- group of the central carbon of citrate in your right hand, and the OH group of the central carbon in your left hand. The CH_2-CO_2^- that came from acetyl-CoA will be pointing up while the CH_2-CO_2^- that came from oxaloacetate will be pointing down. Aconitase holds onto citrate in the same way, and when it moves the hydroxyl group of citrate, it moves it specifically from the central carbon to the CH_2 group on the bottom. Aconitase can move the OH group only because the catalytic apparatus is in the right place. If we were to try to bind citrate to aconitase with the CH_2-CO_2^- that came from acetyl-CoA in the down position, the CO_2^- on the central carbon would be on the left, and this wouldn't do. Aconitase is set up to expect the CO_2^- group on the central carbon to be on the right.

As we follow the label from citrate to succinyl-CoA, there are no problems. But at succinate, it all seems to fall apart.

$$*CH_2CO_2^- \qquad\qquad \longrightarrow \qquad\qquad *CH_2CO_2^-$$
$$| \qquad\qquad\qquad\qquad\qquad\qquad\qquad\quad |$$
$$CH_2C(\!=\!O)SCoA \qquad\qquad\qquad\qquad CH_2CO_2^-$$

Unlike the CH_2-CO_2^- groups of citrate, the two CH_2-CO_2^- groups of succinate are indistinguishable. If you pick up succinate by the labeled CH_2 group, you will always be able to pick up succinate by the unlabeled CH_2 group so that it looks exactly the same (except for the label).

$$
\begin{array}{ccccc}
CO_2^- & & CO_2^- & & CO_2^- \\
| & & | & & | \\
H\!-\!{}^*C\!-\!H & & H\!-\!C\!-\!H & & H\!-\!{}^*C\!-\!OH \\
| & = & | & \longrightarrow & | \\
H\!-\!C\!-\!H & & H\!-\!{}^*C\!-\!H & & H\!-\!{}^*C\!-\!OH \\
| & & | & & | \\
CO_2^- & & CO_2^- & & CO_2^-
\end{array}
$$

Malate

The consequence of the symmetry of succinate is that the enzyme succinate dehydrogenase has a fifty-fifty chance of picking up succinate in either of the equivalent orientations. By the time you get to malate, in which all the carbons are obviously different, one-half of the original label is found at the CH_2 group of malate while the other half of the label is on the CH–OH group of malate. Each molecule of malate has only one

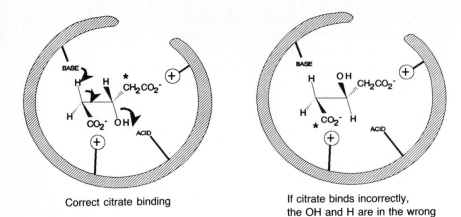

Correct citrate binding   If citrate binds incorrectly,
                                         the OH and H are in the wrong
                                         orientation for the reaction

Figure 20-1
THE TWO $CH_2CO_2^-$ ARMS OF CITRATE ARE NOT IDENTICAL, and the
enzyme aconitase can tell them apart. The $CH_2CO_2^-$ arm that was derived from
acetyl-CoA (*) is not metabolized to CO_2 during the first turn of the TCA cycle.

labeled carbon; however, in the collection of all malate molecules, the
label is equally distributed between the two carbons in the middle [if it
started out on the CH_3 group of acetyl-CoA (C-2)]. The same thing
happens with the label that comes into the TCA cycle from the C-1 of
acetyl-CoA except that the label ends up equally distributed between the
two carboxyl groups of the malate.

Since neither of the carbons that come in from acetyl-CoA is lost
during the first turn of the TCA cycle, it's reasonable to wonder when
they are lost. If the label was originally at C-1 (the C=O) of acetyl-CoA, it
ends up in the two carboxylate groups of oxaloacetate. On the next turn of
the TCA cycle (go around again without bringing any more label in from
acetyl-CoA) both of these carboxyl groups are lost as CO_2. So when C-1
of acetyl-CoA is labeled, all the label is lost from the TCA-cycle intermediates on the second turn of the cycle.

It's a lot more complicated when C-2 of the acetyl-CoA is labeled.
After the first turn of the cycle, this label ends up on the central carbons
of oxaloacetate, and neither of these is lost during the second turn of the
cycle. However, on the third turn of the cycle, half the label is lost
because half the total label is on the carboxylates of oxaloacetate because
of the symmetry of succinate. On each subsequent turn of the cycle, half
the remaining label is lost. The only way to sort this out for yourself is to
sit down with the TCA cycle and go round and round. It's a dizzying
experience that leaves you a little bit nauseated when it's over.

pH, pK_a, pROBLEMS

·

The usual concerns about acid–base behavior (other than knowing in your soul that it really can't be very important) are where protons go if they're given a choice, how buffers work, and logarithms. Your professor's job is to try to convince you that acid–base behavior has a place in biochemistry. Your job is to learn it—just in case your professor is right.

PROTON: H^+ OR H_3O^+

A hydrogen nucleus without electrons

It's interesting that the simplest molecule in biology has succeeded in terrifying generations of chemistry and biochemistry students.

ACID

Something that has a proton it's willing to give up

BASE

Something that has a place to put a proton

The thing that may be confusing here is that when an acid gives up a proton, it becomes a base, and when a base picks up a proton, it becomes an acid. It's a constant identity crisis. To make it even worse, some molecules with protons aren't acids. Every proton could be an acid if the base removing it were strong enough. In water (which limits the strength of acids and bases that can be used) RNH_3^+ is an acid but RNH_2 is not. The difference between the protons in RNH_3^+ and RNH_2 is that RNH_2 is such a weak acid that the strongest base available in water (hydroxide, ^-OH) is too weak a base to remove it (to give RNH^-). On the other hand, RNH_3^+ is a sufficiently strong acid that ^-OH can easily remove one proton to give RNH_2. Whether or not something is an acid (for our purposes) comes down to whether or not the strongest base available in water (^-OH) can remove a proton from it. The same sort of reasoning can be applied to a base. A base (for our purposes) is something that can be protonated by the strongest acid in water (H_3O^+).

There's no easy, foolproof way to decide whether something is an acid or a base. Fortunately, there are only a few types of acids and bases you will encounter in biochemistry. Notice that all bases will be more negatively charged than the acids they came from.

ACID		BASE		pK$_a$
Asp, Glu R—CO—OH Carboxylic acid	$+ H_2O \rightleftharpoons$	R—CO—O$^-$ Carboxylate	$+ H_3O^+$	3–5
Lys, His, Arg (*not* NH of peptide bond) R—NH$_3^+$ Protonated amine	$+ H_2O \rightleftharpoons$	RNH$_2$ Amine	$+ H_3O^+$	9–10 (amines) 6–7 (His)
Cys R—SH Thiol	$+ H_2O \rightleftharpoons$	R—S$^-$ Thiolate	$+ H_3O^+$	8–9
Tyr (*not* Ser or Thr) R—OH Phenol	$+ H_2O \rightleftharpoons$	R—O$^-$ Phenolate	$+ H_3O^+$	9–11
CO$_2$ Carbon dioxide	$+ H_2O \rightleftharpoons$	HCO$_3^-$ Bicarbonate	$+ H_3O^+$	6.1
H$_3$PO$_4$ H$_2$PO$_4^-$ HPO$_4^{2-}$ Phosphoric Acid	$+ H_2O \rightleftharpoons$ $+ H_2O \rightleftharpoons$ $+ H_2O \rightleftharpoons$	H$_2$PO HPO$_4^{2-}$ PO$_4^{3-}$ Phospate	$+ H_3O^+$ $+ H_3O^+$ $+ H_3O^+$	1.9 6.8 10.5
HCl Hydrochloric acid	$+ H_2O \rightleftharpoons$	Cl$^-$ Chloride	$+ H_3O^+$	-4
H$_3$O$^+$ The proton	$+ H_2O \rightleftharpoons$	H$_2$O Water	$+ H_3O^+$	-1.6
H$_2$O Water	$+ H_2O \rightleftharpoons$	HO$^-$ Hydroxide	$+ H_3O^+$	15.8

NOT ALL ACIDS AND BASES ARE CREATED EQUAL

Strong acids completely dissociate in water; weak acids don't.
Strong acids = high K_a = low pK_a.

There are acids and then there are ACIDS. Acids like HCl and H_2SO_4 are strong acids. When you add a mole of HCl to water, you get a mole of protons (H_3O^+) and a mole of Cl^-. HCl and other strong acids are, in fact, stronger acids than H_3O^+ so that the equilibrium

$$HCl + H_2O \rightleftharpoons H_3O^+ + Cl^-$$

is far to the right. For a weaker acid, like acetic acid, the equilibrium

$$HOAc + H_2O \rightleftharpoons H_3O^+ + {}^-OAc$$

is far to the left. This means that when you add acetic acid to water, most of it stays as acetic acid—it doesn't dissociate very much at all.

The strength of an acid can be defined by the tendency of the acid to give up its proton to water. The stronger the acid, the larger the equilibrium constant for the reaction

$$HA + H_2O \rightleftharpoons A^- + H_3O^+$$

$$K_a = \frac{[A^-][H_3O^+]}{[HA]}$$

For strong acids, this equilibrium constant is greater than 1, and for weak acids, K_a is much less than 1. For most acids you find in biochemistry, K_a is much less than 1 and proton transfer from HA to water is not very favorable (nor complete).

$$pK_a = -\log(K_a)$$

The lower the pK_a, the stronger the acid.

The pK_a is one of those unfortunate concepts that turns everything around. It also puts acid strengths on a logarithmic scale like earthquakes and other natural disasters. Logarithms will be discussed later. For right now, just remember that stronger acids have lower pK_a's.

WEAK ACIDS MAKE STRONG BASES (AND VICE VERSA)

The conjugate base of a strong acid is a weak base. The conjugate acid of a strong base is a weak acid.

If something is a strong acid, it naturally gives up a proton easily, and the base that results must not want it back. The conjugate base of a strong acid is, then, a weak base. For the same reason, the conjugate acids of strong bases are weak acids. The word *conjugate* is just thrown in to make it sound like real chemistry. The K_a's can also be used to describe base strengths. Bases don't have K_a's, but their conjugate acids do. A strong acid (high K_a) has a weak conjugate base, and a weak acid (low K_a) has a strong conjugate base.

WHO GETS THE PROTON?

The weakest acid (strongest base)

What happens when you mix an acid and base, or, worse still, two acids and a base—Who gets the proton? Often there's no choice. If there's only one base present (other than water), the base gets the proton. If you add an acid to a mixture of bases, the stronger base takes the proton first. Because we only discuss acid strengths, the stronger base is the one that comes from the weakest acid (has the highest pK_a).

$$RNH_2 \qquad RS^-$$
$$pK_a = 10.5 \qquad pK_a = 8.6$$

Here, when you add an acid, the amino group, the strongest base, gets the proton first. If you add a base to a mixture of acids, the strongest acid gives up protons first.

$$RCO_2H \qquad RNH_3^+$$
$$pK_a = 4.5 \qquad pK_a = 10.5$$

When titrating acids with a strong base, the acids titrate in order of increasing pK_a. A carboxylic acid titrates before a protonated amine when strong base is added.

DON'T FORGET STOICHIOMETRY

When a weak acid is neutralized by a strong base, the concentration of the acid (HA) decreases by the same amount [in molar units (M)] as the concentration of the conjugate base increases.

When you add a strong acid to a solution of a base, the base is converted to an acid. There's not really much to remember here, but it still seems to be confusing to a lot of people. In the example in Fig. 21-1, the acetate that is converted to acetic acid when HCl is added has to come from somewhere (acetate) and has to go somewhere (acetic acid). Not only does the concentration of acetic acid increase, but the concentration of the acetate must *decrease*.

THE SADISTIC LITTLE p

$$pH = -\log_{10}[H^+] = \log_{10}\frac{1}{[H^+]}$$

H^+ must be in M units

The little p is what screws it all up. In an attempt to avoid writing lots of 0's and the $-$ sign, some physical chemist in prehistory condemned whole generations to backward thinking. Perhaps it's what little p stands for that is so confusing: p means, "Take the negative logarithm (base 10) of the next thing that follows." $pH = -\log[H^+]$. This has the effect of making very little numbers (like 0.000000001, or 10^{-9}) much bigger, that is, 9. The smaller the number, the bigger the p of that number is. The little p makes things backward—remember *backward*.

TAKING $\log_{10}(x)$

$$\log(10^x) = x$$
$$\log 1000 = \log(10^3) = 3$$
$$\log 0.001 = \log(10^{-3}) = -3$$

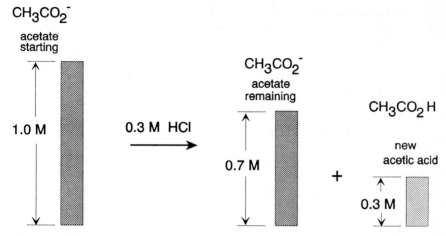

Figure 21-1   A Strong Acid Converts a Base to its Acid Form
Don't forget that the amount of the acid form that's produced has to come from
somewhere.

Logarithms are the answers you get when you put a number in
your calculator and press the $\log_{10}$ button (not to be confused with the ln
button). The $\log_{10}$ of something is the power to which 10 has to be raised
to give the number you just entered: $\log(10^{\text{something}})$ = something, or
$10^{\log(\text{somethingelse})}$ = somethingelse. To undo what log does, use the 10^x
button on your calculator. The logarithm of 1 is zero (10^0 = 1). If a
number is greater than 1, its logarithm is greater than zero (positive). If a
number is less than 1, the logarithm is less than zero (negative). If a
number is negative, then it won't have a logarithm—your calculator
knows this, so don't worry about it.

TAKING $-\log_{10}(x)$

$$-\log (10^x) = -x$$
$$-\log 1000 = -\log (10^3) = -3$$
$$-\log 0.001 = -\log (10^{-3}) = 3$$

or $\quad -\log 0.001 = \log \dfrac{1}{0.001} = \log 1000 = \log (10^3) = 3$

Taking $-\log_{10}$ of something is the same as taking the $\log_{10}$ and then
changing the sign of the answer. There's another way to do this. Enter the
number, take the reciprocal ($1/x$ button), and then take the log (don't
change the sign of anything). Taking the $-\log_{10}$ is the same as taking the

$\log_{10}$ of the reciprocal: $-\log_{10} x = \log_{10} (1/x)$. This is what turns every-thing upside-down when dealing with pH, pK_a, and so forth.

$$\mathbf{pH} = -\mathbf{log_{10}} \, [\mathbf{H^+}]$$

$[H^+] = 10^{-pH}$

Low pH means high $[H^+]$.

Because pH is a little p function, pH and $[H^+]$ are related in a backward manner. The lower the pH, the higher the $[H^+]$. To get pH when you know $[H^+]$, enter the $[H^+]$ (this may require that you use the scientific notation feature on your calculator), press log, and change the sign from $-$ to $+$ (you can do this part in your head). To calculate the $[H^+]$ from the pH, enter the pH on your calculator, change the sign from $+$ to $-$ ($+/-$ button), and hit 10^x. The pH's you'll be dealing with will usually be in the range of 0 to 14, so the $[H^+]$ will be between 1 M (10^{-0}) and $1 \times 10^{-14} \, M$. For pH's that are integers or $[H^+]$'s that are integral powers of 10, you should be able to do it in your head.

$$\mathbf{p}K_a = -\mathbf{log_{10}}(K_a)$$

Lower pK_a = stronger acid, weaker base.

Like pH, pK_a is a "little p" function. Again this turns things back-ward. The lower the pK_a (higher K_a), the stronger the acid.

BUFFERS

Buffers are solutions that contain both the acidic and the basic forms of a weak acid.

Buffers minimize changes in pH when strong acids and bases are added.

A buffer's job is to keep the pH of a solution from changing very much when either strong acids or strong bases are added to it. To handle the addition of both acids and bases, a buffer contains an acid (to react with added base) and a base (to react with added acid). Take a glass of water (*beaker* sounds too chemical, and *blood* sounds too gruesome) and add enough strong acid (HCl) to make the final concentration 10^{-3} *M*. The pH will change from 7 to 3 (Fig. 21-2). Now do the same thing but with a solution of 0.1 *M* KH$_2$PO$_4$ (acid) and 0.1 *M* K$_2$HPO$_4$ (base) (Fig. 21-3). The pH changes from 6.8 to 6.79.

In this example, the starting buffer solution is a little more compli- cated than usual,[1] because in a phosphate buffer there are two bases present—H$_2$PO$_4^{2-}$ and HPO$_4^{-}$. The pK$_a$ leading to H$_2$PO$_4^{-}$ is about 1.9 (H$_3$PO$_4$ $\rightleftharpoons$ H$_2$PO$_4^{-}$ + H$^+$), but the pK$_a$ for the ionization leading to HPO$_4^{2-}$ is 6.8 (H$_2$PO$_4^{-}$ $\rightleftharpoons$ HPO$_4^{2-}$ + H$^+$). So HPO$_4^{2-}$ is a stronger base than H$_2$PO$_4^{-}$ and gets the proton from HCl. When HPO$_4^{2-}$ gets the proton, it turns into H$_2$PO$_4^{-}$. The HPO$_4^{2-}$ concentration then drops by 0.001 *M*, and the concentration of H$_2$PO$_4^{-}$ increases by the same amount. What we're assuming when we do this is that all the protons added with the strong acid end up on the buffer and not in solution. This is basically true. In our example, the initial pH was 6.8, so that [H$^+$] = $10^{-6.8}$ = 1.58 $\times$ 10^{-7} *M*. The final pH was 6.79 ([H$^+$] = $10^{6.79}$ = 1.62 $\times$ 10^{-7} *M*). Even though we added 0.001 *M* HCl, we only generated 0.04 $\times$ 10^{-7} *M* [H$^+$]— 99.96 percent of the protons added ended up on the strongest buffer base and not in the water. This is really the secret of how buffers work; the base and acid forms of the buffer soak up the acids and bases that are added, and very little H$^+$ escapes to change the pH.

HENDERSON-HASSELBALCH EQUATION

$$pH = pK_a + \log \frac{[base]}{[acid]}$$

All concentrations must be in molar units (*M*).
[base] is the concentration of the form of the buffer without the proton.
[acid] is the concentration of the form of the buffer with the proton.
    [base]>[acid] means pH>pK$_a$
    [base]=[acid] means pH=pK$_a$
    [base]<[acid] means pH<pK$_a$

This is one of the two equations you might have to deal with in introductory biochemistry. Like the other equation (the Michaelis-Men- ten equation), this one took two people to derive (wonder which one was

---

[1] If an acetate (CH$_3$CO$_2^{-}$) and acetic acid (CH$_3$CO$_2$H) buffer had been used, the only base present would be acetate, and it would have to get the proton.

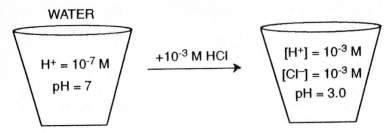

Figure 21-2
WITHOUT BUFFERING, adding a little bit of a strong acid to water decreases the pH a lot.

responsible for deciding to take $-\log$?).[2] If you know pH and pK_a for any buffer, you can calculate [base]/[acid]. If you know pK_a and [base]/[acid], then you can calculate pH. Given two of the three, you can find the other.

The key to understanding the behavior of buffers is in the ratio [base]/[acid]. The ratio [base]/[acid] = 1.0 represents a special point (one that's easy to remember). When [base]/[acid] = 1.0, log ([base]/[acid]) = 0 and pH = pK_a. If [base]/[acid] >1, then pH must be bigger than the pK_a. If [base]/[acid] < 1, the pH must be lower than the pK_a. You may actually be able to get by without memorizing the equation if you realize the general relationships. Just write some version of it down and then fix the signs so that it works right. If your version doesn't give pH>pK_a when [base]/[acid]>1, then just change the sign of one thing and it will work.

TITRATION CURVES

Buffers buffer best within 2 pH units of the pK_a.

Using the Henderson-Hasselbalch equation in a semiquantitative way can help you understand why buffers buffer best when the pH is near the pK_a of the buffer. When the pH is far away from the pK_a, the ratio [base]/[acid] is either very large or very small (depending on whether the pH is greater than or less than the pK_a). When the ratio [base]/[acid] is large or small, the pH is far from the pK_a (for a pH that is 2 units away from the pK_a, the ratio [base]/[acid] is 100 or 0.01). At these extremes, the buffer is about 99 percent base or 99 percent acid (depending on whether

[2] To get the Henderson-Hasselbalch equation, start with $K_a = [H^+][A^-]/[HA]$ and take $-\log$ of both sides: $-\log(K_a) = -\log[H^+] - \log([A^-]/[HA])$, or $pK_a = pH - \log([A^-]/[HA])$, or $pH = pK_a + \log([base]/[acid])$.

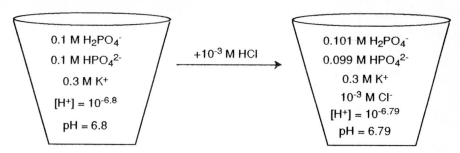

Figure 21-3
WITH BUFFERING, adding a little bit of a strong acid to a buffer doesn't change the pH very much.

pH is above or below the pK$_a$). Adding a little bit of strong acid or base causes the ratio [base]/[acid] to change a lot if the ratio is very large or very small. However, when the ratio is near 1, adding a little bit of strong acid or base has less of an effect on the ratio.

Now let's do it numerically. Let's assume the total concentration of a carboxylic acid is 0.1 M. When we start the titration, the group with the lowest pK$_a$ will start with a [base]/[acid] of 99, [base] = 0.001 M, acid = 0.099 M, and the pH will be about 2 units lower than the lowest pK$_a$ (log 99 is about 2). The amount of dissociation of the carboxylic acid can actually be calculated, but nobody will ask you how to do it. But, remember that it won't dissociate much. Now add enough strong base to make a final concentration of 0.01 M. This will convert 0.01 M of the carboxylic acid to the carboxylate (base) form, resulting in 0.089 M acid and 0.011 M base. The [base]/[acid] ratio will change from about 100 to about 10, and the pH will change by 1 unit. (I'm saying 0.089/0.011 is about 10.)[3] Now let's try the same thing at the pK$_a$. At the pK$_a$, the ratio [base]/[acid] will be 1, [base] = 0.05 M, and [acid] = 0.05 M. When we add 0.01 M base, we will convert 0.01 mole of acid to base; the [base] = 0.06, and the [acid] = 0.04 M. The [base]/[acid] ratio will have changed from 1 to 1.5 (a pH change of 0.18 units). When [base]/[acid] is near 1 (near the pK$_a$), the buffer will buffer best (Fig. 21-4).

With multiple ionizable groups, such as in amino acids and proteins, each group titrates separately according to its pK$_a$. The titration curves shown in Fig. 21-5 are for the amino acids glycine, histidine, and glutamate.

---

[3] I balance my checkbook using the same method, and most of the time it works well, but be aware that there can be some hazards when the account balance approaches the size of the checks that you write.

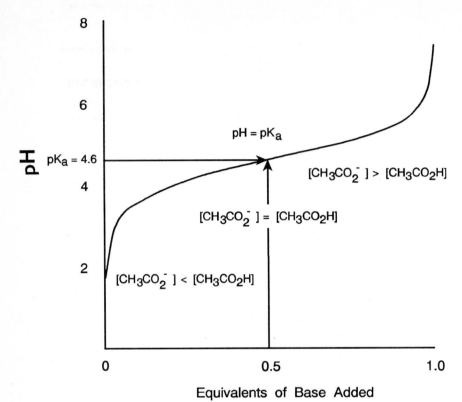

Figure 21-4   Titration of Acetic Acid
At the pK_a, the concentrations of acetate and acetic acid are equal.

In the titration curves shown in Fig. 21-5, you start with the fully protonated form of the amino acid. Notice that at pH's that are not near the pK_a of any functional group, the pH changes more when base is added. Also notice that there are multiple buffer regions (where the pH doesn't change rapidly when base is added) when there are multiple acid and base groups present. When the pK_a's of two groups are close to each other (closer than about 2 units), the higher-pK_a group starts titrating before the lower-pK_a group finishes, and there are no sharp transitions between the two titrations. This is most obvious for the titration of gluta-mate, aspartate, or lysine, which all have two groups with similar pK_a's.

The presence of nearby charged groups can effect the pK_a of an ionizing group. For example, the pK_a of acetic acid (CH_3CO_2H) is 4.6. However, the pK_a of the carboxyl group in glycine ($^+NH_3$–CH_2–CO_2H) is 2.4. The reason lies in the effect of the protonated amino group. The positive charge of this group makes it easier to abstract the proton from the carboxylic acid since the negative charge of the carboxylate product is

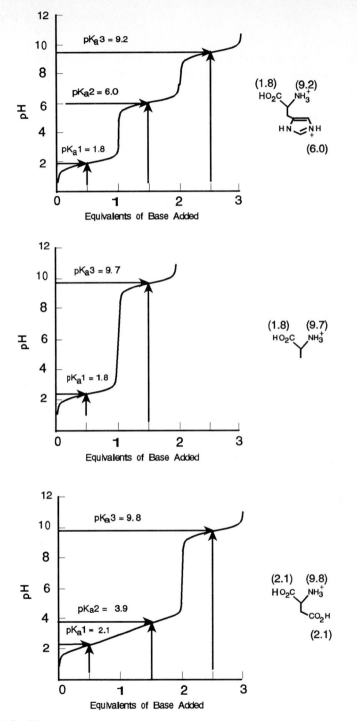

Figure 21-5 Titration of Amino Acids
Multiple ionizable groups titrate separately and in order of increasing pK_a.

stabilized by having the positive charge of the amino group nearby. The effects of nearby charge can be even more dramatic when the ionizable side chain is part of a protein.

---

pI—ISOELECTRIC POINT

pH = pI means no net charge.
pH > pI means negative charge.
pH < pI means positive charge.

The pI is a fake little-p function and does not represent the $-\log[I]$. It represents the pH at which a molecule has no net charge. When fully protonated, glycine has a net $+1$ charge, a charged amino group, and an uncharged, protonated carboxylic acid group. About halfway between the titration of the carboxyl and the amino group, the molecule must have no net charge—the amount of positive charge on RNH_3^+ is exactly matched by the amount of negative charge on $RCOO^-$. If an amino acid (or protein) has more basic groups (Arg, Lys, His) than acidic groups (Glu, Asp), the pI is greater than 7. If an amino acid or protein has more acidic groups, the pI is less than 7. The same considerations apply (but in a more complicated way) to proteins (Fig. 21-6).

At the pI, the molecule is neutral. $[H^+]$ is positively charged, and adding it to another molecule must make that molecule more positively charged. As the pH decreases below the pI, the amount of positive charge on the molecule increases (protonating a $-COO^-$ makes the compound more positive, just like protonating an RNH_2 group). When the pH is less than the pI, the molecule is positively charged. When the pH is greater than the pI, the molecule must be negatively charged.

THE BICARBONATE BUFFER

$$[H^+] = 24\ pCO_2/[HCO_3^-]$$

The CO_2–bicarbonate buffer is a little different from buffers using the usual kind of acids and bases, but it is extremely important in maintaining the acid–base balance of the blood. The acid form of the bicarbonate buffer is actually a gas dissolved in water. Dissolved CO_2 is turned into an acid by hydration to give H_2CO_3. Hydrated CO_2 is then much like a carboxylic acid. It gives up a proton to a base and makes bicarbonate, HCO_3^-.

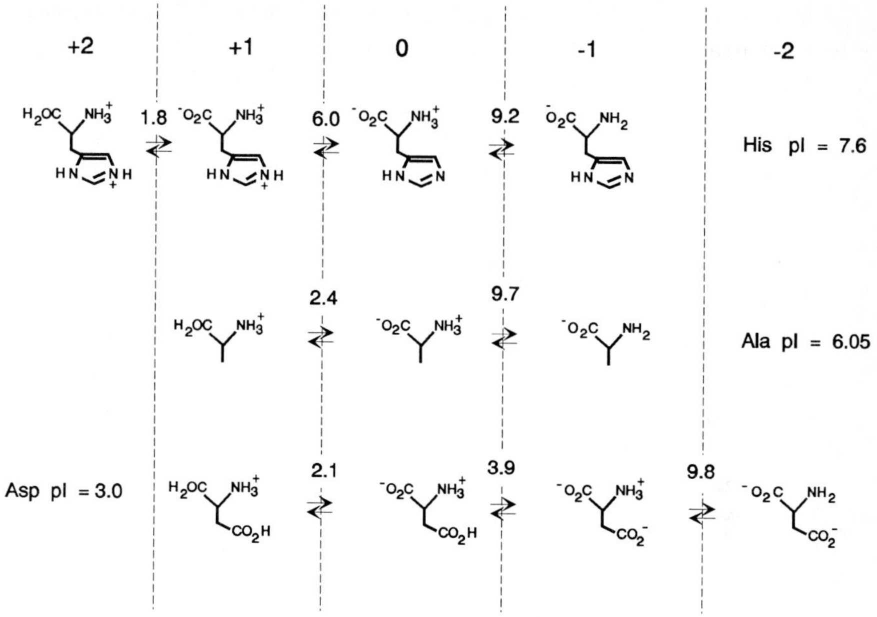

+2	+1	0	-1	-2

H_2OC NH_3^+ 1.8 ^-O_2C NH_3^+ 6.0 ^-O_2C NH_3^+ 9.2 ^-O_2C NH_2

His pI = 7.6

2.4   9.7

H_2OC NH_3^+ ^-O_2C NH_3^+ ^-O_2C NH_2

Ala pI = 6.05

Asp pI = 3.0   2.1   3.9   9.8

H_2OC NH_3^+ CO_2H ^-O_2C NH_3^+ CO_2H ^-O_2C NH_3^+ CO_2^- ^-O_2C NH_2 CO_2^-

Figure 21-6   Figuring out the pI
Average the two pK$_a$'s that convert the charge from +1 to 0 and from 0 to −1.
This will give you an estimate of the pI.

$$CO_2 + H_2O \rightleftharpoons H_2CO_3 \rightleftharpoons H^+ + HCO_3^-$$

When CO_2 is dissolved in water, there is never very much H_2CO_3, so we can ignore it and count CO_2 as the acid and HCO_3^- as the base.

There are two ways of dealing with the bicarbonate buffer system. The first uses the Henderson-Hasselbalch equation and an effective pK$_a$ of 6.1. If there is more base (HCO_3^-) than acid (CO_2), the pH will always be bigger than the pK$_a$. This is usually the case physiologically (pH = 7.4; pK$_a$ = 6.1) so that on a molar basis there is always more than 10-fold more HCO_3^- than CO_2.

You might be wondering why the bicarbonate buffer can buffer effectively at pH 7.4 when its pK$_a$ is 6.1. The answer is that it doesn't buffer all that well. What makes it unique and the major buffer system of the blood is that CO_2, being a gas, can be exhaled by the lungs. Exhaling CO_2 is equivalent to exhaling protons.

$$H^+ + HCO_3^- \rightleftharpoons CO_2 + H_2O$$

It's not that a proton is exhaled; it's just left behind and turned into water.

This gives the body control over the concentration of the CO_2 by controlling the breathing rate.

The other way to think about the CO_2–bicarbonate buffer is *not* to use logarithms. One reason this can be done is that the pH of the body doesn't change all that much (at least under conditions compatible with life), and so you don't have to deal with very small numbers. The penalty for not using logarithms is that you have to worry about the units of things. The concentration of CO_2, being a gas, is usually expressed in terms of the partial pressure of CO_2 in the gas above a liquid containing a given concentration of CO_2 (sounds obscure, doesn't it?). The CO_2 "concentration" is generally expressed in pressure units of millimeters of mercury and is denoted pCO_2. This is not a little-p function. The little p just stands for pressure (as in pO_2). A typical CO_2 partial pressure is about 40 mmHg. Atmospheric pressure is 760 mmHg. This means that the CO_2 represents 40/760 of the total pressure of the gas above the liquid, or about 5%. The actual concentration of dissolved CO_2 (dCO_2; in millimoles per liter) is given by

$$dCO_2 \quad (mmol/L, \text{ or } mM) = 0.03\ pCO_2 \quad (mmHg)$$

Starting the same place as the Henderson-Hasselbalch equation

$$K_a = \frac{[H^+][dCO_2]}{[HCO_3^-]}$$

we know that $K_a = 10^{-6.1} = 7.94 \times 10^{-7}\ M$, and that $dCO_2 = 0.03\ pCO_2$. Putting these things together and rearranging things results in a bit of magic:

$$[H^+] \quad (neq/L) = \frac{24\ pCO_2 \quad (mmHg)}{[HCO_3^-] \quad (mM)}$$

Here's the magic part. Divide pCO_2 in millimeters of mercury by $[HCO_3^-]$ in millimoles per liter (mM) and multiply by 24 (2 dozen). The answer you get is the $[H^+]$ concentration in units of nanoequivalents per liter (neq/L) ($nM = 10^{-9}\ M$). An equivalent (equiv) in the bicarbonate system is the same as a mole, and the terms neq/L, nmol/L, nM, and 10^{-9} M all mean the same thing.

To get $[H^+]$ in molar units (M) (that's what is needed to determine pH), multiply the $[H^+]$ in neq/L by 10^{-9}. Take the $-\log$ of this and you have pH. Whichever way you use, the important thing to remember is that when pCO_2 increases, the $[H^+]$ increases (the pH decreases)—after all, CO_2 is an acid.

IMBALANCE IN BLOOD pH

CO_2 concentration is regulated by the lungs.
$[HCO_3^-]$ is regulated by the kidneys.

By changing the breathing rate and depth, the lungs can control the concentration of CO_2 in the blood. CO_2 is a gas, and when it leaves the body it essentially turns protons into water. The trick is that the actual acid form of this buffer, H_2CO_3 (which you get by adding protons to HCO_3^-), can lose water (not protons—water) and give CO_2, which can then leave the body. This process doesn't leave the protons behind, it just converts them into water. By breathing fast and deeply (hyperventilating), one causes more CO_2 to leave the body than normal, and the pCO_2 in the blood drops. Shallow and slow breathing (hypoventilation) causes the blood pCO_2 to rise.

The kidney exerts control over the concentration of bicarbonate. The initial filtrate of the kidney has the same buffer composition (except for the proteins) as serum. Based on the pH of the initial filtrate, the kidney can decide to reclaim the initially filtered bicarbonate and place it back in the serum, or it can decide to just let it go out of the body. If the kidney does not reclaim bicarbonate, the serum bicarbonate concentration falls. If the kidney reclaims more bicarbonate than normal, the bicarbonate concentration rises.

Normally, everything is balanced, CO_2 and bicarbonate are removed at exactly the same rate that they're formed, and the pH of the serum remains very close to 7.4. If for some reason the lungs blow off too much CO_2 by hyperventilating (breathing too fast), the numerator of the magic equation (pCO_2) decreases, and, as long as $[HCO_3^-]$ remains constant, the $[H^+]$ must decrease (pH increases).

$$[H^+]\downarrow \ = \ \frac{24pCO_2 \downarrow}{[HCO_3^-]}$$

Hypoventilating (breathing too slowly) increases the pCO_2 (numerator of the magic equation), and, as long as $[HCO_3^-]$ remains constant, the $[H^+]$ must increase (pH decreases).

$$[H^+]\uparrow \ = \ \frac{24pCO_2 \uparrow}{[HCO_3^-]}$$

If you just remember that CO_2 is an acid, it's easy to see that if you increase the amount of the acid component of the buffer, the pH must fall.

Kidneys can secrete $[H^+]$ (by an energy-dependent proton pump) directly into the filtrate, lowering the pH of the urine. Because of the lowered pH, more of the initially filtered bicarbonate exists as the gas, CO_2, which can freely cross membranes. Diffusion of this acidic gas back to the blood (through the epithelial cells) and hydration and loss of the proton to form bicarbonate results in reclaiming the initially filtered bicarbonate. The kidney reclaims bicarbonate by actually reclaiming the acid form (CO_2), and the net effect is reclaiming bicarbonate and a proton. The excess protons must actually exit the body on another buffer in the urine (usually phosphate). The net result is that as the pH of the urine falls, the kidney reclaims more and more of the bicarbonate. Reclaiming more HCO_3^- in response to a decreased serum pH ($[H^+]\uparrow$) causes the $[HCO_3^-]$ to rise, restoring the pH to normal.

$$[H^+] = \frac{24pCO_2 \uparrow}{[HCO_3^-] \uparrow}$$

The kidneys, by reclaiming bicarbonate, can compensate for an increased pCO_2 caused by a problem with the lungs.

ACIDOSIS AND ALKALOSIS

Acidosis: pH < 7.4
Alkalosis: pH > 7.4
 Respiratory acidosis: $pCO_2 \uparrow$
 Respiratory alkalosis: $pCO_2 \downarrow$
 Metabolic acidosis: $[HCO_3^-] \downarrow$
 Metabolic alkalosis: $[HCO_3^-] \uparrow$

Normally everything is in balance. The lungs have the numerator (pCO_2), and the kidneys have the denominator ($[HCO_3^-]$). The amount of CO_2 produced by metabolism is balanced by the amount of CO_2 blown off by the lungs and let go by the kidney. It's the same way with the other metabolic acids—they leave the body as CO_2 or they are excreted by the kidneys. If everything always stayed in balance, you wouldn't have to learn all this—but as usual it doesn't, and you do.

When blood pH is not normal, something must not be working correctly or the capacity of the system must have been exceeded. Using the magic equation

$$[H^+] = \frac{24pCO_2}{[HCO_3^-]}$$

we can see that there are two things that might cause the [H$^+$] to be abnormal: a change in pCO$_2$ or a change in [HCO$_3^-$]. If the [H$^+$] is abnormally high (low pH), the condition is called *acidosis*. If the [H$^+$] is abnormally low, the condition is called *alkalosis*.

For each type of [H$^+$] imbalance (too high or too low), there are two possible causes. An abnormally high [H$^+$] could have been caused either by an abnormally high pCO$_2$ (increased numerator of the magic equation) or by an abnormally low [HCO$_3^-$] (decreased denominator). Since pCO$_2$ is regulated by the lungs, the label *respiratory* is attached to effects that cause the pCO$_2$ to change. An acid–base imbalance caused by a change in pCO$_2$ is termed *respiratory acidosis* or *alkalosis* depending on whether the pCO$_2$ is increased or decreased. The bicarbonate concentration is the province of the kidney and is determined by how much bicarbonate the kidney reclaims. Changes in the bicarbonate concentrations are labeled *metabolic*.

When something goes wrong with acid–base balance, something is not working right. If the lungs are at fault and the pCO$_2$ changes, it is called *respiratory acidosis* if pCO$_2$ increases ([H$^+$] increases) or *respiratory alkalosis* if the pCO$_2$ decreases ([H$^+$] decreases). When the kidneys are at fault, and [HCO$_3^-$] is too high or low, it's called *metabolic alkalosis* or *acidosis*. If the bicarbonate concentration drops, then its called metabolic acidosis, and if the bicarbonate concentration rises, it's called metabolic alkalosis because the change in [H$^+$] must be opposite that of the bicarbonate (look at the magic equation again). Thus, we have four possible causes of acid–base imbalance: respiratory acidosis, respiratory alkalosis, metabolic acidosis, and metabolic alkalosis. The simple thing to remember is cause (metabolic or respiratory) and direction (acidosis or alkalosis).

The body has two options when acid–base balance breaks down—fix the original problem or do something else to provide a temporary fix until a more permanent fix can be implemented. If a person hypoventilates (breathes out too little CO$_2$), blood pCO$_2$ rises and the pH decreases ([H$^+$] increases). This is respiratory acidosis. Obviously, if the body could increase respiration, there wouldn't be a problem. Since that doesn't seem to be an option, the next strategy would be to change the bicarbonate concentration to get the pH back to normal. Which direction should it go? Look at the magic equation. If pCO$_2$ is increased (numerator), the only way to get the [H$^+$] back to normal is also to increase [HCO$_3^-$]—metabolic alkalosis.

$$[H^+] = \frac{24pCO_2 \uparrow}{[HCO_3^-] \uparrow}$$

The conclusion—respiratory acidosis must be compensated by metabolic alkalosis in order to return the pH to normal. The table below summarizes all the possibilities. Note that the directions of the arrows are always the same for pCO_2 and $[HCO_3^-]$. Metabolic is always compensated by respiratory (and vice versa), and acidosis is always compensated by alkalosis.

CAUSE		RESULT		COMPENSATION
Respiratory acidosis	$pCO_2 \uparrow$	$[H^+] \uparrow$	$[HCO_3^-] \uparrow$	Metabaolic alkalosis
Respiratory alkalosis	$pCO_2 \downarrow$	$[H^+] \downarrow$	$[HCO_3^-] \downarrow$	Metabolic acidosis
Metabolic acidosis	$[HCO_3^-] \downarrow$	$[H^+] \uparrow$	$pCO_2 \downarrow$	Respiratory alkalosis
Metabolic alkalosis	$[HCO_3^-] \uparrow$	$[H^+] \downarrow$	$pCO_2 \uparrow$	Respiratory acidosis

THERMODYNAMICS AND KINETICS

·

Thermodynamics

Free Energy

Adding Free-Energy Changes

Coupling Free Energies

Thermodynamic Cycles

$\Delta G = \Delta H - T\Delta S$

Driving Force

Kinetics

Velocity

Transition State Theory

Rate Constants

Rate Constants and Mechanism

· · · · · · · · · · · · ·

THERMODYNAMICS
Alarms sound, eyes go blank, and a sigh can be heard.

The beauty of thermodynamics is that it can tell you whether or not a chemical reaction can occur and how much energy you can get out of it when it does. The beast of thermodynamics is that no one who really understands it can (or will) explain it to those of us who don't.

Take a generic chemical reaction:

$$A + B \rightleftharpoons P + Q$$

What will happen if we mix A, B, P, and Q together? There's some gray area here in that the answer depends somewhat on what we mean by *happen*. First, it depends on direction. A more appropriate way to ask the question is, Will the reaction happen in the direction written, that is, left to right? Second, it depends on the actual concentrations of A, B, P, and Q that you start with. Third, it really depends on the relationship between the initial concentrations of A, B, P, and Q and the equilibrium concentrations that will exist when the reaction finally comes to equilibrium. Finally, when A, B, P, and Q are mixed, they will take off toward the equilibrium position, whatever that is, but thermodynamics doesn't tell you how long it mght take for the reaction to actually get to equilibrium. The *How fast?* is kinetics. So the real answer is that when we mix A, B, P, and Q, the reaction will happen in the direction that takes you to equilibrium. When the reaction is actually at equilibrium, the concentrations of A, B, P, and Q will be equal to their equilibrium concentrations.

The equilibrium constant for a reaction is just the ratio of the products to the reactants at equilibrium:

$$K_{eq} = \frac{[P]_{eq}[Q]_{eq}}{[A]_{eq}[B]_{eq}}$$

If the initial ratio of products to reactants,

$$\frac{[P][Q]}{[A][B]}$$

is different from the equilibrium ratio, the chemical reaction will proceed until the real product/reactant ratio equals the equilibrium product/substrate ratio, and then it stops at equilibrium.

If $([P][Q]/[A][B]) < ([P]_{eq}[Q]_{eq}/[A]_{eq}[B]_{eq})$, the reaction goes in the direction that increases P and Q and decreases A and B so that the product/substrate ratio increases to the equilibrium value. This is the same as saying that if the concentration of products is lower than their equilibrium values, the reaction goes in the direction that makes more products, or to the right.

If $([P][Q]/[A][B]) > ([P]_{eq}[Q]_{eq}/[A]_{eq}[B]_{eq})$, the reaction goes in the direction that decreases P and Q and increases A and B so that the product/reactant ratio decreases to the equilibrium value. The reaction goes to the left.

If $([P][Q]/[A][B]) = ([P]_{eq}[Q]_{eq}/[A]_{eq}[B]_{eq})$, the reaction is at equilibrium and there will be no net change in the products or substrates.

FREE ENERGY

ΔG is a measure of how far a chemical reaction is from equilibrium.
ΔG represents the amount of work that can be done by a chemical
    reaction.
You get more useful work (ΔG) out of a chemical reaction if it is far
    from equilibrium.

	$\Delta G = 0$	Reaction is at equilibrium, and the concentrations of products and reactants equal the equilibrium concentrations.
Exergonic reaction:	$\Delta G < 0$	Reaction happens in the direction written. Reactant concentrations are *higher* than the equilibrium concentration.
Endergonic reaction:	$\Delta G > 0$	Reaction happens in the direction opposite to that written. Product concentrations are higher than the equilibrium concentration.

$$\Delta G = \Delta G^0 + RT \ln \frac{[P][Q]}{[A][B]}$$
$$\Delta G^0 = -RT \ln (K_{eq})$$
$$\Delta G^0 = -1.36 \log_{10} (K_{eq})$$

A 10-fold difference in K_{eq} or the products/reactants ratio changes
ΔG by 1.36 kcal/mol.

Another way to think about this involves the energies of the reactants and products. Chemical reactions occur in the direction written when the products of the reaction have less energy than the reactants. The reaction proceeds from a state of higher energy to one of lower energy—from a mountain to a valley. These are called exergonic reactions because they give off energy. Reactions don't proceed uphill, from a valley to a mountain—unless there's a way to couple the unfavorable, uphill movement to a more favorable, downhill movement of something else.

The energy we're talking about is called the free energy G. The energy is not free as in no charge; it's free like a bird. The free energy is the energy that is available to use. The free energy of a molecule is something that cannot be measured; however, we can measure the change in free energy (ΔG) that accompanies a chemical reaction.

If the products have less G than the reactants (the reaction is downhill), the ΔG is less than zero ($\Delta G = G_{products} - G_{reactants}$). Thus, for a spontaneous reaction (one that occurs in the direction written), the ΔG is negative (<0). The more negative ΔG, the more favorable the reaction. Like most things biochemists do, this too initially seems backward (Fig. 22-1).

Just by looking at the value of ΔG, you can determine which way a reaction goes. If $\Delta G < 0$, the reaction goes to the right. If $\Delta G > 0$, the reaction goes to the left. And, if $\Delta G = 0$, the products and reactants are of exactly the same free energy (note that this does not mean that the products and reactants are at the same concentration), and the reaction is at equilibrium.

What's important in determining how much free energy is available from a given chemical reaction is how far the reaction is from its equilibrium position. One way to decide this is to take the ratio of the product/reactant ratio at equilibrium to the actual product/reactant ratio:

$$\frac{[P]_{eq}[Q]_{eq}/[A]_{eq}[B]_{eq}}{[P][Q]/[A][B]} = \frac{K_{eq}}{[P][Q]/[A][B]}$$

If this ratio is 1, the reactants and products are at equilibrium. If this ratio is large, the reaction has more reactants present (or less products) than at equilibrium and the reaction will go to the right, toward the products. The opposite is true if the ratio is small—the reaction will go toward the reactants.

Now, the definition—because that's all it really is. The ΔG is the negative natural logarithm ($-\ln$) of the ratio of the equilibrium constant to the products/reactants ratio multiplied by the absolute temperature T in degrees Kelvin, and the gas constant R in calories per degree per mole. The reason for taking the natural logarithm is to take a simple, easily understandable ratio and make it difficult. The multiplication by the temperature and gas constant is to give it units of energy (Calories/mol or kcal/mol).

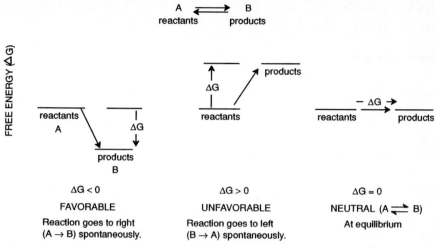

Figure 22-1 Reactions Don't Go Uphill

$$\Delta G = -RT \ln \frac{K_{eq}}{[P][Q]/[A][B]}$$

or we can rearrange this to give

$$\Delta G = -RT \ln K_{eq} + RT \ln \frac{[P][Q]}{[A][B]}$$

ΔG *is only a measure of how far a chemical reaction is from equilibrum.* We also can make another definition

$$\Delta G^0 = -RT \ln K_{eq}$$

ΔG^0 is the free-energy change for a reaction under conditions where the product/reactant ratio is 1.[1] Don't get confused on this point—ΔG^0 is *not* the free-energy change at equilibrium (that's zero), it's the free energy change when the products/reactants ratio is 1. ΔG^0 is a way to compare different reactions to decide which one is intrinsically more favorable. The comparison is made, by convention, at a product/reactant ratio of 1. Just because a reaction has a negative ΔG^0 doesn't mean that it can't be

---

[1] The product/reactant ratio may have units if there are more product terms than reactant terms, or vice versa. For example, if there are two products and one reactant, the product/reactant ratio will have molar units (*M*). In this case, a products/reactants ratio of 1 means that the products/reactants ratio is actually 1 *M*. The term *molar standard state* means that we're talking about a products/reactants ratio that has molar units.

made to go in the reverse direction. If the actual product/reactant ratio is large enough to overcome the ΔG^0, the reaction can be made to go in the reverse direction. The ΔG^0 is just a convenient way to compare reactions under standard conditions. The position of some reactions also depends on temperature and pH. Since biochemists usually work at pH 7.0 and 25°C, pH and temperature effects can be ignored by adding a prime to ΔG^0 and calling it $\Delta G^{0\prime}$. This means that this is the ΔG^0 at pH 7.0 and 25°C.

When all the algebra is done,

$$\Delta G = \Delta G^0 + RT \ln \frac{[P][Q]}{[A][B]}$$

For the most part, equations don't tell most people much. But this one is different. It tells everybody exactly how to decide what's going to happen. There are just two things (and only two things) that go into this decision. First, there's how much ΔG is built into the system—certain chemical reactions are just intrinsically more favorable. This factor shows up in ΔG^0. Second, there's the size of the actual products/reactants ratio. If it's large enough, it can overcome even a large intrinsic free-energy difference.

To make it a little simpler to calculate ΔG from real numbers, it's useful to remember that natural logarithms (ln) can be converted to base-10 logarithms (log),[2] and that at 25°C, the value of RT is 0.591 kcal/mol. To calculate ΔG^0, for example, use the equation

$$\Delta G^0 = -1.36 \log_{10} (K_{eq})$$

For a K_{eq} of 1×10^4 (an intrinsically favorable reaction), $\log_{10} K_{eq} = 4$, and $\Delta G^0 = -1.36(4) = -5.44$ kcal/mol. For a K_{eq} of 1×10^{-6} (an intrinsically unfavorable reaction), $\log_{10} K_{eq} = -6$, and $\Delta G^0 = -1.36(-6) = 8.16$ kcal/mol.

• **UNITS** The products/reactants ratios may have units associated with them. For example, a reaction of the type $A \rightleftharpoons B + C$ has a products/reactants ratio that has molar units. What you do when you take the log of a products/reactants ratio with molar units is ignore the units. You've not really made them disappear, you've just ignored them. The way physical chemist types make this difficult is that they call ignoring the units an *assumption of standard state*. It does matter, though. If you

[2] $\ln x = 2.303 \times \log_{10} x$.

assume the units are molar (M), the products/reactants ratio has one value, but if you assume the units are millimolar (mM), or micromolar (μM), the products/reactants ratio can be very different. A molar standard state is different from a millimolar standard state. You must remember the units that have been used (even though you've just ignored them) and make sure that both K_{eq} and the products/reactants ratio are expressed in the same units.

Let's do a real example. The hydrolysis of ATP provides energy for virtually everything alive. The K_{eq} for this reaction (ATP $\rightleftharpoons$ ADP + P_i) at 25°C and pH 7.0 is about 2.3×10^5 M. Notice that the concentration of water has also been ignored here. That's because it's large and relatively constant in biological systems, and putting it in would just change the numbers for everything by the same amount—so it was decided to ignore it.

With a K_{eq} of 2.3×10^5 M, the $\Delta G^{0\prime}$ is $-1.36 \log_{10} (2.3 \times 10^5) = -7.3$ kcal/mol. This would be the ΔG if the products/reactants ratio were 1, but it's not. Let's assume that the local concentration of ATP in a cell is 5 mM, [ADP] is 60 μM, and [P_i] is 5 mM (these are approximately right, but they will vary from cell to cell and at any given time in a cell). Keep in mind that this also means the amount of free energy available from ATP hydrolysis will vary from cell to cell and from time to time.

With these concentrations, the products/reactants ratio becomes

$$\frac{[ADP][P_i]}{[ATP]} = \frac{(6 \times 10^{-5}\ M)\ (5 \times 10^{-3}\ M)}{(5 \times 10^{-3}\ M)}$$

or products/reactants = 6×10^{-5} M. Putting all those things into the equation for ΔG,

$$\Delta G = \Delta G^0 + RT \ln \frac{[P][Q]}{[A][B]}$$

$$\Delta G = \Delta G^0 + 1.36 \log_{10} (6 \times 10^{-5})$$

With $\Delta G^0 = -7.3$ kcal/mol

$$\Delta G = -7.3\ \text{kcal/mol} - 5.7\ \text{kcal/mol} = -13\ \text{kcal/mol}$$

Because there's so much more ATP than ADP in most cells, we can get more free energy (by -5.7 kcal/mol) out of ATP hydrolysis under physiological conditions than you would think just by looking at the value of $\Delta G^{0\prime}$

ADDING FREE-ENERGY CHANGES

Free-energy changes for chemical reactions can be added and subtracted to give free-energy changes for other chemical reactions.

Free-energy changes, like some of the other thermodynamic properties, can be added and subtracted—sometimes to let you evaluate a chemical reaction that can't actually be observed. The reason these additions and subtractions of free-energy changes are possible is that the free-energy change observed for a chemical reaction doesn't depend at all on how the overall reaction was accomplished. The jargon is that free-energy changes are *path-independent*. All that matters is *that* you get from point A to point B, not *how*.

ATP hydrolysis can be accomplished directly:

$$ATP + H_2O \rightleftharpoons ADP + P_i \qquad \Delta G^{0\prime} = -7.3$$

or ATP can be hydrolyzed by the following series of chemical reactions:

$$\begin{aligned}
ATP + \text{F-6-P} &\rightleftharpoons ADP + \text{F-1,6-P}_2 & \Delta G^{0\prime} &= -4.6 \text{ kcal/mol} \\
\text{F-1,6-P}_2 + H_2O &\rightleftharpoons \text{F-6-P} + P_i & \Delta G^{0\prime} &= -2.7 \text{ kcal/mol} \\
\hline
\text{Sum: } ATP + H_2O &\rightleftharpoons ADP + P_i & \Delta G^{0\prime} &= -7.3 \text{ kcal/mol}
\end{aligned}$$

If you want to prove to yourself that these free energies sum, you can write the equilibrium expressions for the first and second reactions and multiply them together, and you'll get the equilibrium expression for the hydrolysis of ATP. Multiplication is equivalent to the addition of logarithms, so that when you multiply equilibrium constants, you're actually adding free energies (or vice versa).

COUPLING FREE ENERGIES

The free energy of a favorable chemical reaction can be used to make an unfavorable reaction happen.

The formation of a peptide bond (as in proteins) is not a favorable reaction. Hydrolysis of the peptide bond would be the spontaneous reaction:

$$AA_1\text{—}CO_2^- + {}^+NH_3\text{—}AA_2 \rightleftharpoons$$
$$AA_1\text{—}CONH\text{—}AA_2 + H_2O \quad \Delta G^{0\prime} = +0.5 \text{ kcal/mol}$$

If we guess that the intracellular concentrations of the amino acids are 1 mM each for AA_1 and AA_2, and the dipeptide concentration is also 1 mM, the ΔG would actually be $+4.6$ kcal/mol. This tells you that peptide bonds *cannot* be made under these conditions.

Biologically, the unfavorable formation of a peptide bond is driven by the hydrolysis of ATP and GTP so that the reaction can actually happen:

$$AA_1\text{—}CO_2^- + ATP + tRNA \rightleftharpoons AA_1\text{—}C(\!=\!O)\text{—}O\text{—}tRNA + PP_i$$

$$AA_1\text{—}COO\text{—}tRNA + 2GTP + {}^+NH_3\text{—}AA_2 \rightleftharpoons$$
$$AA_1\text{—}CONH\text{—}AA_2 + tRNA + 2GDP + 2P_i$$

The reaction is pushed even further to completion by the hydrolysis of the pyrophosphate from the first reaction:

$$PP_i + H_2O \rightleftharpoons 2P_i$$

Sum: $AA_1 + AA_2 + ATP + 2GTP \rightleftharpoons$
$$AA_1\text{—}AA_2 + AMP + 2GDP + 4P_i$$

The formation of a peptide bond with an estimated ΔG^0 of $+4.6$ kcal/mol is driven by coupling the process to the hydrolysis of 4 high-energy phosphate bonds [estimated ΔG^0 of $4 \times (-7.3 \text{ kcal/mol})$]. The overall ΔG^0 for the process would then be ($+4.6$ kcal/mol $-$ 29.2 kcal/mol $= -24.6$ kcal/mol). There are numerous examples of the coupling of ATP hydrolysis to otherwise unfavorable reactions. Prime examples are the transport of ions from a compartment that has a low concentration of the ion to a compartment that has a high concentration of the ion. The movement of an ion (or other molecule) against a concentration gradient (from low concentration to high concentration) is not thermodynamically favorable. Ion pumps that concentrate ions against a concentration gradient require either the hydrolysis of ATP or the simultaneous transport of an ion down its concentration gradient. The sole reason for all these couplings of free energy is that if they weren't coupled, the reactions wouldn't happen. Nothing happens if the free-energy change is greater than zero.

THERMODYNAMIC CYCLES

The sum of the free-energy change around any cyclic path must be zero.

The product of equilibrium constants around any cyclic path of reactions must equal 1.

The additivity of free-energy changes can be extended by the idea that free energy doesn't depend on the pathway a chemical reaction takes, just on the identity of the products and the reactants.

Think about a reaction between a small molecule and a protein in which the small molecule (ligand) binds to the protein, forming a protein–ligand complex. Now, let's also suppose that when the protein has ligand bound to the active site, the conformation of the protein is different from when ligand is not bound. We'll call P^* the form of the protein with the conformation that is found when the ligand is bound. If we were actually to measure an equilibrium constant for the association reaction (K_{obs}) or a ΔG^0 for the association reaction, we would be measuring the ΔG^0 for a reaction of a protein P with a ligand L to give a protein–ligand complex in which the structure of the protein was different from the structure when ligand was not bound:

$$P + L \xrightleftharpoons{K_{obs}} P^*L$$

But we can think about this process in two different ways. First, we could say that the ligand binds to the protein P and that this causes the protein to change its conformation to P^*:

$$P + L \xrightleftharpoons{K_1} PL \xrightleftharpoons{K_2} P^*L$$

Alternatively, we could say that the protein exists in two forms, P and P^*, even in the absence of ligand. The ligand picks out the form of the protein P^* and binds to it:

$$P + L \xrightleftharpoons{K_3} P^* + L \xrightleftharpoons{K_4} P^*L$$

Without any more information, we can't say which of the two possibilities actually happens. The nice thing about thermodynamics is that it doesn't matter. Now let's see if you can be convinced of this. We know we can add free energies of individual reactions to get the free-energy change of another reaction.

$$
\begin{array}{llll}
\text{P + L} \rightleftharpoons \text{PL} & \Delta G_1 & \text{or} & K_1 \\
\text{PL} \rightleftharpoons \text{P}^*\text{L} & \Delta G_2 & \text{or} & K_2 \\
\hline
\text{Sum: P + L} \rightleftharpoons \text{P}^*\text{L} & \Delta G_1 + \Delta G_2 & \text{or} & K_1 K_2
\end{array}
$$

Now do the other path:

$$
\begin{array}{llll}
\text{P + L} \rightleftharpoons \text{P}^* + \text{L} & \Delta G_3 & \text{or} & K_3 \\
\text{P}^* + \text{L} \rightleftharpoons \text{P}^*\text{L} & \Delta G_4 & \text{or} & K_4 \\
\hline
\text{Sum: P + L} \rightleftharpoons \text{P}^*\text{L} & \Delta G_3 + \Delta G_4 & \text{or} & K_3 K_4
\end{array}
$$

Hopefully, you will have noticed by now that when these two "different" pathways are followed, the overall reaction that has been accomplished is the same and is equal to the reaction we originally called K_{obs}. (If you haven't noticed this, then notice it now—it's the point of this whole section.)

What we've just shown is that

$$K_{obs} = K_3 K_4 = K_1 K_2$$

or

$$\Delta G_{obs} = \Delta G_1 + \Delta G_2 = \Delta G_3 + \Delta G_4$$

Another way of saying the algebra in words is that regardless of how we go from P + L to P^*L, the free-energy change is the same. The ΔG for all paths is the same—so are the equilibrium constants. Free-energy change is independent of path. We can also show this on what is called a *thermodynamic box*. The top and right of Fig. 22-2 are one pathway, the left and bottom are another, and the diagonal is a third. For any thermodynamic box, the sum of the free-energy changes going all the way around the box must be zero—you're starting and stopping in the same place. Since the sign of ΔG depends on direction, if you go through a step backward to the way it was defined, you need to change the sign of ΔG for that step or take the reciproal of the equilibrium constant. Going completely around the box above using the equilibrium constants gives

$$K_1 K_2 \frac{1}{K_3} \frac{1}{K_4} = \frac{K_1 \, K_2}{K_3 \, K_4} = 1$$

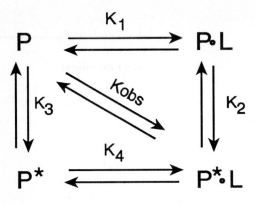

Figure 22-2
A THERMODYNAMIC
CYCLE describing a change
in the conformation of a
protein that accompanies the
binding of a substrate.

The utility of thermodynamic boxes lies in using them as thinking tools—particularly in cases in which there are alternative ways of thinking about a particular chemical process. The above example for a protein structural change coupled to the binding of a ligand illustrates the point. Along the top and right of the box, we would argue that the ligand bound to the protein and caused the conformation change in the protein. By the left and bottom pathway we would argue that the ligand just trapped one of two normal conformations of the protein and pulled the equilibrium toward the ligand–protein complex. The two pathways along the edges of the box give different pictures of reality, both of which are thermodynamically equivalent. The other take-home message is that without additional experimental data, you can't really decide between the two alternative mechanisms. What the box tells you is that before you can really decide which mechanism is correct, you've got to find evidence for PL (top pathway) or P^* (bottom pathway).

$$\Delta G = \Delta H - T\Delta S$$

ΔG = free energy (<0 for favorable reactions). This is the useful energy that can be obtained from a chemical reaction.

ΔH = enthalpy. This is the net amount of energy available from changes in bonding between reactants and products. If heat is given off, the reaction is favorable and $\Delta H < 0$.

ΔS = entropy. This is the change in the amount of order during a reaction. Order is unfavorable ($\Delta S < 0$). Disorder is favorable ($\Delta S > 0$).

The free-energy change for a chemical reaction, ΔG, is a balance between two factors—heat and organization. Other things being equal, reactions that give off heat are more favorable than those that don't. Reactions that make more disordered products also tend to be more favorable than reactions that make more organized products. Heat energy arises from chemical reactions by making and breaking chemical bonds. But not all the heat generated from a chemical reaction can be used. Some of the heat energy may have to be used to organize or order the products of the reaction.

• ΔH (ENTHALPY CHANGE)   A reaction gives off or takes up heat because of changes in the chemical bonding that accompany the reaction. This can amount to forming more bonds in the products than in the reactants, or it can mean that the bonds in the reactants are more energetic than the bonds in the products. Reactions that give off heat (called *exothermic*) form more stable bonds in the products than in the reactants and tend to be more favorable than reactions that don't give off heat. A ΔH (enthalpy change) that is negative (heat given off) makes ΔG more negative and favors the reaction in the direction written. Heat can be considered just like a reactant or product in a chemical reaction. You can even write it down if you like.

• ΔS (ENTROPY)   Most chemical reactions are also accompanied by a change in the organization of the reactants and products. For example, the protein-folding reaction takes a structureless, random protein and converts it into a folded and well-organized three-dimensional structure. The structure has become organized, and organization is unfavorable—it doesn't happen spontaneously. *Entropy* is the word given to disorder (the opposite of organized). High-entropy systems are disorganized, whereas low-entropy systems are organized. As before, it's products minus reactants. If the products are more organized (low entropy) than the reactants (high entropy), the ΔS is negative, but the contribution to the free energy, which is $-T\Delta S$, is positive—unfavorable. If organization accompanies a chemical reaction, it makes an unfavorable contribution.

DRIVING FORCE

The factor contributing the most to the negative ΔG. The driving force makes it happen.

Driving force is a term that is used to describe what provides most of the favorable free-energy change for a chemical reaction, that is, what makes it happen. We know that the ΔG for a chemical reaction that happens in the direction written must be less than zero. If, for example, there is a chemical reaction in which the net enthalpy change (ΔH) is zero or positive and the reaction still happens, we know that the entropy contribution must be in the favorable direction (and positive). Otherwise, ΔG couldn't be negative. In this case, we would say that the reaction is entropically driven (i.e., the reaction occurs because of the increase in entropy, increased disorder, that accompanies the reaction).

Driving force can also be used to denote the physical interaction that provides the most negative free energy to an overall reaction. For example, the driving force for the folding of proteins is the hydrophobic interaction. Other forces contribute to a favorable protein-folding reaction; however, the largest contribution comes from the packing of the hydrophobic residues into the interior of the protein.

To make it more meaningful, let's take the reaction for the formation of a hydrogen bond in water:

$$—C{=}O\cdots HOH + H_2O\cdots H—N—R \rightleftharpoons$$
$$R—C{=}O\cdots HN—R + H_2O\cdots HOH$$

The reactants, on the left, are a carbonyl oxygen atom that, in the absence of any other hydrogen-bond donor, forms a hydrogen bond to water, and a hydrogen from an amide that is also hydrogen-bonded to water. The reaction involves bringing the $C{=}O$ and HN– groups together, breaking the two hydrogen bonds to water, and then forming a new hydrogen bond between the $C{=}O$ and HN– groups and a new hydrogen bond involving the two water molecules. Will this reaction happen?

Let's look at the enthalpy change. We have two hydrogen bonds on the left and two hydrogen bonds on the right. They're between different species, but the overall reaction is just a rearrangement of hydrogen bonds without actually changing the number. One would think that the ΔH for this reaction should be near zero, and it is. But if we're talking about a hydrogen bond forming in a protein in which the carbonyl oxygen and the amide nitrogen are positioned at a good distance and angle for hydrogen-bond formation, or if the hydrogen-bond donor or acceptor is charged, one of the hydrogen bonds that's formed could be a little bit stronger than the ones involving water. The ΔH could be a little bit negative.

Now try entropy. On the left side of the reaction, the hydrogen-bond donors and acceptors are free to move through three-dimensional space independently. They are free. There is some restriction, however, because they are interacting with a water molecule. But water molecules are everywhere, and only a small number of the total water molecules have restricted motion. The left side of the reaction is reasonably disorganized. The right side of the reaction has the peptide hydrogen-bond donor and acceptor forming a hydrogen-bonded complex. In this complex, the donor and acceptor must move through three-dimensional space together (that's what a complex is). Each molecule has lost some of its freedom to move independently (each can still vibrate and rotate some internal bonds). The two hydrogen-bonded water molecules are just plain water—comparable in organizational state to the hydrogen bonds on the left side of the reaction. Altogether, the right side of the reaction is considerably more organized than the left. The reaction, then, is accompanied by a decrease in entropy (an organization)—an unfavorable proposition.

In balance, the small decrease in enthalpy ($\Delta H < 0$) is more than offset by a large decrease in entropy ($\Delta S < 0$) so that the overall reaction is unfavorable. Thus, one would not expect to see the formation of single hydrogen bonds between two peptides in water. This is what is found.

In proteins and some small peptides, one does see hydrogen-bond formation. Intramolecular hydrogen-bond formation is not as entropically unfavorable as intermolecular hydrogen-bond formation, so that intramolecular hydrogen bonds (as in an α helix) are more likely to form on entropic grounds. In addition, hydrogen-bond formation in DNA and proteins is cooperative. It may be hard (entropically) to form the first hydrogen bond, but after it's formed it becomes easier and easier to form additional hydrogen bonds—less and less entropy loss is required for forming additional hydrogen bonds.

KINETICS

Alarms sound faster, eyes go blank quickly, and a long sigh can be heard.

Thermodynamics tells you *what* will happen—given enough time. Kinetics supplies the *when*. Don't get too bogged down by all of this. Look back at the trivia sorter to put it in perspective. However, if you really want to at least partially understand what kinetics is about, proceed.

VELOCITY

How fast
Change in concentration with time
M/min

Kinetics is all about change with time.

$$A \longrightarrow P$$

For a simple chemical reaction like the conversion of A to P we can ask, How fast? *How fast* is measured in terms of velocity—the change in the concentration of substrate or product with time.

$$\text{Velocity} = v = \frac{\Delta[P]}{\Delta\text{time}}$$

Velocity, v, has units of M/min.

TRANSITION STATE THEORY

Free energy of activation: Energy required to raise the energy of the reactant to the energy of the transition state

Transition state: Highest-energy arrangement of atoms that occurs between the reactants and product

---

All chemical reactions don't occur with the same velocity; some are faster than others. Before a reaction can happen, the individual molecules of A must have enough thermal energy to break or make the chemical bonds that constitute the chemical reaction. A chemical reaction can be viewed as being blocked by a barrier (Fig. 22-3). This barrier keeps the reaction from happening instantaneously (if it weren't there, we would all be CO_2 and water). If the barrier is low, the reaction is fast, but if the barrier is high, the reaction is slow. The energy that a molecule of A must gain before it will undergo the chemical reaction (cross over the barrier) is called the *free energy of activation* ($\Delta G^{\ddagger}$). The energy (usually) comes from thermal vibration of the molecule. The activation energy determines how fast a given reaction happens.

Transition state theory tells us that when a molecule of substrate has enough energy to jump the barrier, its structure is intermediate between that of the substrate and that of the product. Some bonds are stretched, partially broken, partially formed, and so forth. The arrangement of atoms that has the highest energy between the substrate and product is called the *transition state*. Transition state theory assumes that the transition state doesn't exist for more than the time required for one bond vibration (about 10^{-15} s)—so the transition state really doesn't exist, but we can talk about it as if it did. The ΔG's of activation are always positive. The more positive, the slower.

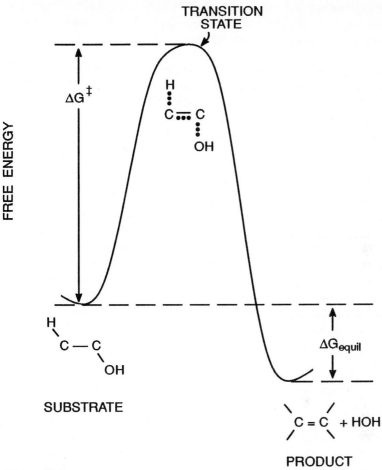

Figure 22-3
A **FREE-ENERGY REACTION COORDINATE DIAGRAM** shows the free energy of the substrate, product, and transition state of a chemical reaction. It tells you how favorable the overall reaction is (ΔG_{eq}) and how fast ($\Delta G^{\ddagger}$).

RATE CONSTANTS

First-order: $v = k\,[A]$
   k with units of time^{-1} (min^{-1} or s^{-1})
Second-order: $v = k\,[A][B]$
   k with units of concentration^{-1} time^{-1} (e.g., M^{-1} s^{-1})
Zero-order: $v = k$
   k with units of concentration time^{-1} (e.g., M s^{-1})

Rate constants are numbers that tell us how fast a reaction happens. They come in a few different flavors and have different units depending on the type of reaction they're describing.

• **FIRST-ORDER REACTIONS**   The velocity, or rate, of a reaction is the change in substrate concentration per unit time. For a simple reaction of the type A → P, the velocity of the formation of P or the disappearance of A is found (usually) to be proportional to the concentration of A that is present at the time the velocity is measured:

$$v = \frac{-d[A]}{dt} = k\,[A]$$

This equation is known as a *rate law*. It tells you how the rate of the reaction depends on the concentration(s) of the substrate. The order of the reaction is defined as the power to which the substrate concentration is raised when it appears in the rate law. In the case above, [A] is raised to the first power ($[A]^1$), so the reaction is said to be first-order with respect to the A concentration, or simply *first-order in A*. The rate constant k is a proportionality constant thrown in so that the equation works and so that the units work out. Since v must have units of molar per second (M/s) and [A] has molar units (M), then k must have units of reciprocal seconds (1/s or s^{-1}).

For reactions that are first-order, the rate of the reaction depends on how much substrate is present at any time ([A]) (Fig. 22-4). As the reaction proceeds, substrate is used up and converted to product. As the substrate is used up, the velocity decreases. The rate is not constant with time. The fraction of A molecules that have sufficient energy to react is constant under a given set of reaction conditions. If 100 out of 1000 molecules have enough thermal energy to react, after the reaction of these molecules has occurred, there will be 90 molecules out of the remaining 900 that have enough energy to react. Since the velocity decreases as the substrate is used up, a plot of [A] against time is a curved line; the slope decreases with time. As you can see, as the reaction proceeds, the slope (velocity) decreases. To actually find out how the concentration of substrate changes with time (so we can actually draw the graph), we've got to do a little bit of calculus (just kidding). Let's just write down the answer:

$$A = A_0 e^{-kt}$$

Here A is the concentration of substrate A at any time t, A_0 is the initial concentration of A at time zero, and k is the first-order rate constant. The

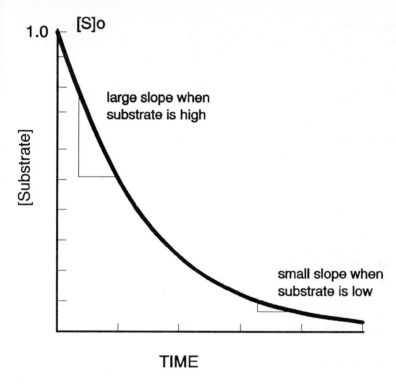

TIME

Figure 22-4
For a **FIRST-ORDER REACTION,** the velocity decreases as the concentration of substrate decreases as it is converted to product. As a result, a plot of substrate concentration against time is a curved line.

concentration of A at any time is an exponential function of time. At short times ($t = 0$), $e^{-kt} = 1$ ($e^0 = 1$), and $A = A_0$.

One consequence of a first-order reaction is that it takes a constant amount of time for half the remaining substrate to be converted to product—regardless of how much of the reactant is present. It takes the same amount of time to convert 100,000 A molecules to 50,000 P molecules as it takes to convert 10 A molecules to 5 P's. A first-order reaction has a constant half-time $t_{1/2}$.

When half the initial amount of A has disappeared,

$$A = 0.5A_0$$

Putting this into the exponential equation,

$$0.5A_0 = A_0\, e^{-kt_{1/2}}$$

Cancelling the A_0,

$$0.5 = e^{-kt_{1/2}}$$

Taking the ln of both sides,

$$\ln 0.5 = -kt_{1/2}$$
$$\ln \tfrac{1}{2} = -kt_{1/2}$$
$$-\ln 2 = -kt_{1/2}$$
$$\ln 2 = kt_{1/2}$$
$$0.692 = kt_{1/2}$$
$$\frac{0.692}{k} = t_{1/2}$$

All that arithmetic is just to show you that the half-time for a first-order reaction depends only on k, not on how much A you have to start with. The whole point is that the bigger the k, the shorter the half-time, the faster the reaction.

An exponential function that describes the increase in product during a first-order reaction looks a lot like a hyperbola that is used to describe Michaelis-Menten enzyme kinetics. It's not. Don't get them confused. If you can't keep them separated in your mind, then just forget all that you've read, jump ship now, and just figure out the Michaelis-Menten description of the velocity of enzyme-catalyzed reaction—it's more important to the beginning biochemistry student anyway.

• **SECOND-ORDER REACTIONS**   For a second-order reaction, the velocity depends on the concentration of two molecules. Reactions of the type

$$A + B \longrightarrow P$$

usually (not always—but don't worry about the exceptions) follow a rate law of the type

$$v = k\,[A][B]$$

where the velocity at any time depends on the concentrations of both the reactants. Change either of them and you change the rate of P formation. The reason is that both A and B have to have enough energy to react with each other and only a fraction of each has the necessary energy.

We say that the reaction is first-order with respect to the concentration of A, first-order with respect to the concentration of B, and second-order overall. The overall reaction order is the sum of all the individual reaction orders of the substrates that appear in the rate law. If the rate law were $v = k[A]^2[B]$, the rate would be second-order in A, first-order in B, and third-order overall. The units of k for a second-order reaction are reciprocal molar seconds (M^{-1} s^{-1}), again to make the units match.

Just as with the first-order case, the velocity (rate) of product appearance (or substrate disappearance) changes as the reaction proceeds. Most of the time, one of the reactants (A or B, it doesn't matter which) will be present in large excess over the other reactant. For example, let's say we start a reaction between A and B in which the A concentration is 5 μM but the B is 0.5 mM (100 times higher than A). At the beginning of the reaction the concentration of B is 0.5 mM; after the reaction is over and all the A is used up (converted to P), the concentration of B is 0.495 mM. Over the entire reaction, B has changed only 1%, an insignificant amount. The concentration of B has remained constant over the entire reaction, and all the changes in velocity that have occurred resulted from the decrease in the A concentration during the reaction. Since $k[B]$ is a constant, the disappearance of A appears first-order (i.e., exponential) and the observed first-order rate constant (k_{obs}) will be given by $k[B_0]$, where k is the second-order rate constant (units of M^{-1} s^{-1}) and $[B_0]$ is the constant concentration of B that is present. A plot of the observed first-order rate constant for the disappearance of A (or appearance of P) against the concentration of B is a straight line whose slope is k, the second-order rate constant (Fig. 22-5).

• ZERO-ORDER REACTIONS   Reactions that are zero-order in everything follow the rate law

$$v = k$$

In contrast to the other reaction orders, the velocity of a zero-order reaction does not change with the concentration of the substrate or with time (Fig. 22-6). The velocity (slope) is a constant and k has the units molar per minute (M/min, or M min^{-1}). Reactions that are zero-order in absolutely everything are rare. However, it is common to have reactions that may be zero-order in the reactant that you happen to be watching. Let's think of a two-step reaction

$$A + B \underset{k_{-1}}{\overset{k_1}{\rightleftharpoons}} \text{Intermediate} \overset{k_2}{\longrightarrow} P$$

where we are measuring the appearance of P. Both k_2 an k_{-1} are first-order while k_1 is second-order, first-order in A and first-order in B.

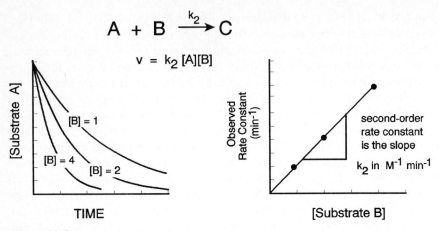

Figure 22-5
The rate of **SECOND-ORDER REACTIONS** depends on the concentration of both substrates. If the concentration of B is constant, A disappears in a first-order fashion, but the rate constant for A disappearance depends on the concentration of B.

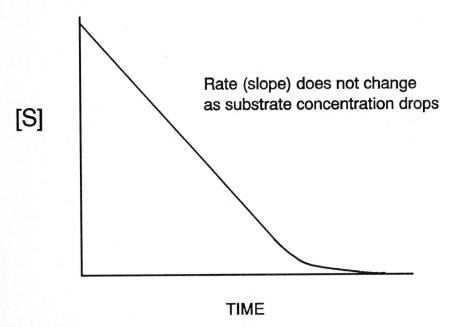

Figure 22-6
The rate of a **ZERO-ORDER REACTION** does not change as the substrate concentration changes. As a result, a plot of substrate concentration against time is a straight line (the velocity is constant with time).

Suppose that the reaction between A and B to give the intermediate is very fast and very favorable. If we have more B than A to start with, all the A is converted instantly into the intermediate. If we're following P, what we observe is the formation of P from the intermediate with the rate constant k_2. If we increase the amount of B, the rate of P formation won't increase as long as there is enough B around to rapidly convert all the A to the intermediate. In this situation, the velocity of P formation is independent of how much B is present. The reaction is zero-order with respect to the concentration of B. This is a special case. Not all reactions that go by this simple mechanism are zero-order in B. It depends on the relative magnitudes of the individual rate constants. At a saturating concentration of substrate, many enzyme-catalyzed reactions are zero-order in substrate concentration; however, they are still first-order in enzyme concentration (see Chap. 7).

RATE CONSTANTS AND MECHANISM

Complex mechanisms can show simple kinetic behavior.

Even relatively complex reactions can behave very simply, and 99 percent of the time, understanding simple first-, second-, and zero-order kinetics is more than good enough. With very complicated mechanistic schemes with multiple intermediates and multiple pathways to the products, the kinetic behavior can get very complicated. But more often than not, even complex mechanisms show simple kinetic behavior. In complex mechanisms, one step (called the *rate-determining* step) is often much slower than all the rest. The kinetics of the slow step then dictates the kinetics of the overall reaction. If the slow step is simple, the overall reaction appears simple.

APPENDIX

CHEMICAL STRUCTURE AND FUNCTIONAL GROUPS

Functional Groups

RNH$_2$	R$_2$C=NH	R-C(=O)NH-R	
amine	imine	amide	
ROH	R-C(=O)-H	R-C(=O)R	R-C(=O)-OH
alcohol	aldehyde	ketone	carboxylic acid

R-C(=O)-OH
ester

RSH	RSR	RSSR	R-(C=O)-SR
thiol	sulfide	disulfide	thiol ester
CH$_3$CH$_3$	CH$_2$=CH$_2$	HC≡CH	
alkane	alkene	alkyne	

$$\overset{\overset{O}{\|}}{HO-P-O^-} \quad \overset{\overset{O}{\|}}{RO-P-O^-} \quad \overset{\overset{O}{\|}}{RO-P-OR} \quad \overset{\overset{O}{\|}}{RO-P-OR}$$

O	O-	O-	OR
phosphate	phospho-monoester	phospho-diester	phospho-triester

OR	OR	OH	OH
R-C-H	R-C-R	R-C-H	R-C-R
OR	OR	OR	OR
acetal	ketal	hemiacetal	hemiketal

256

R-groups

$$-\overset{\overset{\displaystyle O}{\|}}{C}-CH_3$$
acetyl

$$-\overset{\overset{\displaystyle O}{\|}}{C}-R$$
acyl

$-OH$
hydroxyl

$-NH_2$
amino

$-CO_2^-$
carboxyl

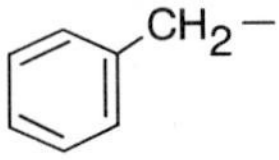

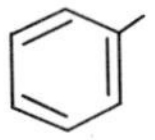

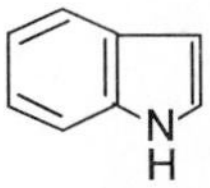

benzyl

phenyl

imidazole

indole

VITAMINS[1]

Name	Structural Feature	Reactions
Biotin	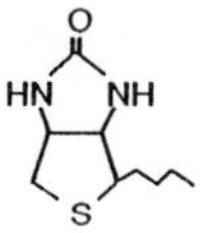	most transfers of CO_2 pyruvate carboxylase
Cobalamins (B_{12})	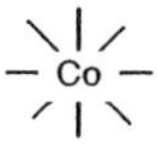 $- Co -$	methylmalonyl-CoA mutase (odd-chain fatty acid metabolism) methionine synthesis
Ascorbic acid	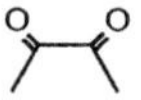 OR 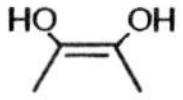	hydroxylation of proline and lysine in collagen/elastin synthesis nor-adrenaline synthesis
Folic acid	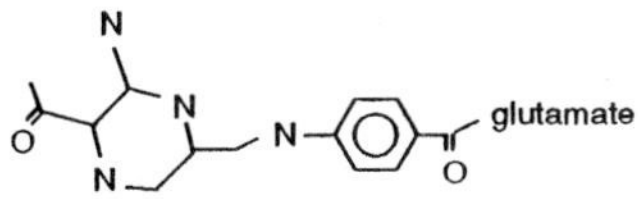 glutamate	deoxy-TMP synthesis from deoxy-UMP; required for DNA synthesis purine and pyrimidine synthesis one-carbon metabolism
Nicotinamide	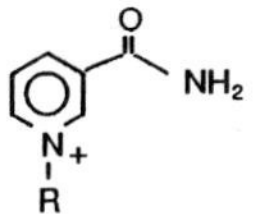 NH_2 OR NH_2 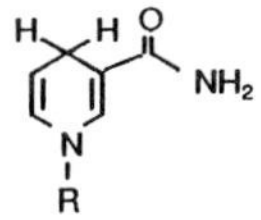	oxidations and reductions throughout metabolism (NADH, NADPH)
Pyridoxal (B_6)	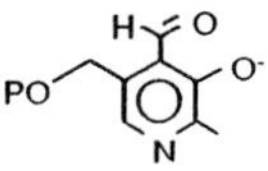	transaminases, decarboxylases, epimerases; almost everything to do with amino acid metabolism

[1] The structures of the vitamins and cofactors shown in this table are not complete strutures. Only the funtional parts of the molecules are shown.

VITAMINS *(Cont.)*

Pantothenate

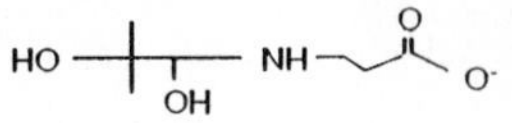

HO ─┬─ NH ─── O⁻
   OH

a part of coenzyme A;
you'll never see it alone

Coenzyme A

BIG STRUCTURE — SH

fatty acid synthesis and oxidation;
carries groups at acyl (acid) oxidation state

(i.e., CoA-S$-\overset{O}{\overset{\|}{C}}$R)

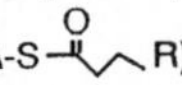

Riboflavin

R

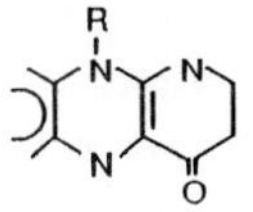

cofactor for oxidation and reduction

succinate dehydrogenase – TCA

Thiamine

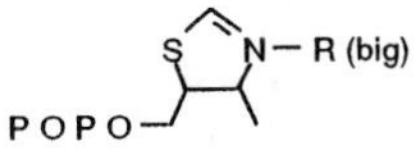

N — R (big)

POPO

transfer of $\overset{O}{\overset{\|}{C}}CH_3$

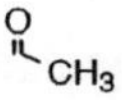

pruvate dehydrogenase
α-ketoglutarate dehydrogenase (TCA)

HMP-pathway – transaldolase and
transketolase

Lipoic acid

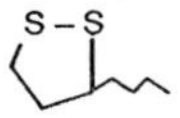

S – S

oxidations and reductions and acyl transfer –
pyruvate dehydrogenase
α-ketoglutarate dehydrogenase

Vitamin A
(retinoic acid)

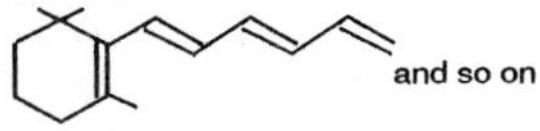

and so on

visual pigment and other, unknown functions

Vitamin E
(α-tocopherol)

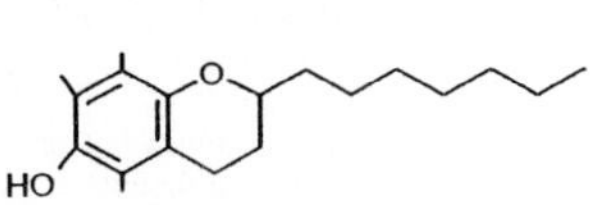

HO

lipid-soluble antioxidant to protect membranes

Vitamin K

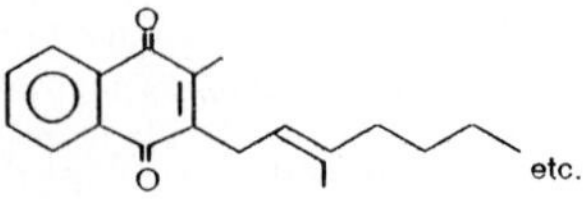

etc.

carboxylation at glutamate residues in proteins of
blood clotting system to create Ca^{2+} binding site

Acetoacetate. $[CH_3C(=O)CH_2CO_2^-]$. **A ketone body.** Produced by the liver during the metabolism of fat. Ketone bodies can be used by other tissues (not the liver itself) for fuel during fasting and starvation. Their synthesis in the liver regenerates CoASH to allow continued fatty acid oxidation.

Acetyl-CoA. $CH_3C(=O)-SCoA$. End product of fatty acid β oxidation and beginning point for fatty acid synthesis. May be made from glucose through pyruvate and pyruvate dehydrogenase; however, acetyl-CoA cannot be converted to glucose or glucose equivalents.

Acetyl-CoA carboxylase. $CH_3C(=O)-SCoA + CO_2 + ATP \rightarrow {}^-O_2CCH_2C(=O)-SCoA$. The enzyme that catalyzes the synthesis of malonyl-CoA, the rate-limiting step in fatty acid synthesis. Requires the cofactor *biotin. Inactivated* by *phosphorylation. Activated* by *citrate.*

Acid. A proton donor. The groups RCOOH, the protonated imidazole of histidine, R-SH, RNH_3^+, and R-OH of tyrosine are the only ones in proteins that are acidic enough to worry about. Acids are always more positively charged than the bases they come from.

Acidic residue. Glu, Asp. Amino acid side chains that are such strong acids that they have already lost their proton at neutral pH. They add negative charge to a protein.

Activation energy. The amount of energy that a molecule or molecules must have gained before they will react chemically. The difference in energy between the substrate and transition state. The more activation energy required, the slower the reaction.

Active site. A specialized region of an enzyme where the enzyme interacts with the substrate and catalyzes its conversion to products. Active sites are saturable—at high enough concentration of substrate, all the enzyme molecules have their active sites occupied.

Acyl. $R-C(=O)-X$. R is any carbon chain, and X is anything except carbon or hydrogen. Usually used in front of something like CoA, as in acyl-CoA $[R-C(=O)-SCoA]$.

Adenosylcobalamin. Vitamin B_{12}. Recognize it by the cobalt atom. Involved in the methylmalonyl-CoA mutase reaction (metabolism of propionyl-CoA from odd-chain-length fatty acids) and in the transfer of a methyl group from methyl-THF to homocysteine to regenerate methionine.

Adenylate cyclase. Makes cAMP. The enzyme responsible for making cAMP out of ATP. It is stimulated by glucagon and epinephrine through a receptor-mediated mechanism involving a G protein.

Adipose tissue. Fat tissue. Responsbile for storing triglyceride. Adipose tissue cannot use glycerol for fuel or for triglyceride synthesis because it lacks the enzyme glycerol kinase. A hormone-sensitive lipase of the adipose tissue hydrolyzes triglyceride and releases the fatty acids in response to cAMP (*glucagon* and *epinephrine*).

Alanine cycle. Alanine exported from muscle is converted to urea and glucose in liver. A cooperative pathway between liver and muscle in which the ammonia and carbon from amino acid metabolism are removed from the muscle as alanine, taken up by the liver, transaminated to pyruvate, converted into glucose, and shipped back out to the muscle.

Aldehyde. R–C(=O)–H. The open-chain (noncyclic) form of many sugars is an aldehyde.

Aldol. R–CH(OH)–CH₂C(=O)–R. The product of the condensation between two ketones (or an aldehyde and ketone). Glyceraldehyde 3-phosphate and dihydroxyacetone phosphate undergo an aldol condensation to form fructose 1,6-bisphosphate (which is an aldol).

Aldose. A sugar that has an aldehyde [R–C(=O)–H] functional group when it is written in the open-chain form. Glucose is an aldose. *See* ketose.

Aliphatic. Not aromatic. Contains a bunch of CH_3– and –CH_2– groups but no aromatic rings.

Alkane. –CH₂– hydrocarbon. A hydrocarbon containing a bunch of CH_3– and –CH_2– groups.

Alkene. –C=C– hydrocarbon. A hydrocarbon containing one or more double bonds.

Alkyl. A functional group (R) that's got a bunch of CH_3– and –CH_2– groups.

Alkyne. C≡C hydrocarbon. A hydrocarbon containing a triple bond.

Allosterism. A type of enzyme regulation in which an effector binds to one site on the enzyme and increases or decreases the activity at another site.

Amide. R–C(=O)–NH–R. Bond used to connect amino acids together in proteins. Also found in the side chains of Gln and Asn.

Amine. RNH_3^+. A basic (as in acid–base) group in proteins. pK_a is 8.5 to 10. Found in Lys and at the amino-terminal end of a protein.

Amphipathic, Amphiphilic. Both hydrophobic and hydrophilic sides. A helix or sheet structure in which one face (or side) is composed of hydrophilic groups and one face (or side) has hydrophobic groups. These structures pack nicely into proteins or lipid bilayers where the hydrophobic side can interact with other hydrophobics and the hydrophilic side can be exposed to water.

Anabolism. Biosynthetic metabolism. Anabolic pathways are induced in times of abundant energy and glucose.

Anaerobic. No oxygen. Without oxygen, all energy must come from anaerobic glycolysis, which produces lactate as a product.

Anneal. Recombine single-stranded DNA into double-stranded molecules. Double-stranded DNA can be denatured to single strands by increasing the temperature. When the temperature is lowered, denatured (single-stranded) DNA will find its complementary strand and form a double-stranded structure. Denaturation (melting) is facilitated by low salt, and annealing is facilitated by high salt.

Antigen. A foreign (nonself) molecule that can be recognized by the immune system. When the immune system recognizes a foreign molecule, it selects and expands the immune cells that recognize (and destroy) the antigen, resulting in an increased serum concentration of antibodies that recognize specific regions (epitopes) of the foreign molecule.

Antimycin. Oxidative phosphorylation inhibitor. Inhibitor of electron transfer from cytochrome b to cytochrome c_1. Site II. Inhibitors stop electron transfer (oxygen utilization), substrate utilization, and ATP synthesis.

Aromatic. Having $2n + 2$ electrons in a cyclic ring system. Phe, Tyr, and Trp are the aromatic amino acids. Note that histidine is also aromatic, but it's not generally called one of the aromatic amino acids.

Assay. Determining how much of an enzyme you have. The act of measuring how fast a given (or unknown) amount of enzyme converts substrate to product. The act of measuring a velocity.

Asymmetric center. A carbon or other atom that lacks symmetry in the distribution of substituents around it. For tetrahedral (four substituents) carbon, the carbon is asymmetric if the four groups are different. The true test of asymmetry is that the molecule and the mirror image of the molecule cannot be superimposed.

ATPase. An enzyme that hydrolyzes ATP. Generally, an ATPase has some other principal activity. Few enzymes just hydrolyze ATP for the fun of it. For example, the F_1F_0 ATPase makes ATP as protons are transferred from the outside of the mitochondria to the inside. However, it will actually hydrolyze ATP if there is no proton gradient.

Atractyloside. An inhibitor of oxidative phosphorylation. Prevents the exchange of ADP for ATP across the mitochondrial membrane catalyzed by the ADP–ATP translocase. Since the ATP made by oxidative phosphorylation can't get out of the mitochondria, this inhibitor effectively stops oxygen consumption and substrate oxidation when the mitochondrial ADP has all been converted to ATP.

B$_{12}$. Vitamin. Also called adenosylcobalamin or cobalamine. Recognize it by the cobalt atom. A B$_{12}$ cofactor is involved in the methylmalonyl-CoA mutase reaction (metabolism of propionyl-CoA from odd-chain-length fatty acids) and in the transfer of the methyl group from methyl-THF to homocysteine to regenerate methionine.

B$_6$. Vitamin. Also called pyridoxal phosphate. B$_6$ is involved in almost all decarboxylations, eliminations, or epimerizations that involve amino acids.

Base. A proton acceptor. The groups RNH_2, the nitrogens of the imidazole of histidine, $R-S^-$, $R-O^-$ of tyrosine, and the $R-C(=O)-O^-$ are the only ones in proteins that are basic enough to worry about. Bases are always more negatively charged than the acids they come from.

Basic residue. Lys, Arg. Amino acid side chains that still have a proton attached at neutral pH. Basic residues add positive charge to a protein (at neutral pH).

Biopterin. Cofactor for phenylalanine hydroxylase. Biopterin is a redox-active cofactor that is required in the tetrahydro form during the conversion of phenylalanine to tyrosine catalyzed by phenylalanine hydroxylase. The dihydrobiopterin produced by the reaction must be reduced by dihydrobiopterin reductase. An absence of this reductase also causes phenylketonuria (PKU).

Biotin. A cofactor involved in carboxylation reactions. Most enzymes that catalyze the ATP-dependent addition of CO_2 to a substrate (like acetyl-CoA carboxylase) require the cofactor biotin.

Bongkrekate. Inhibits ATP–ADP translocase. Inhibitor of oxidative phosphorylation that prevents the exchange of ADP for ATP across the mitochondrial membrane. ADP is converted totally to ATP inside the mitochondria so that respiration and substrate oxidation stop.

Buffer. Solution of an acid and its conjugate base. If acid is added to a buffer, the basic form of the buffer is converted to the acidic form of the buffer. If base is added, the acidic form of the buffer is converted to the basic form of the buffer. If the ratio of basic to acidic form of the buffer is near 1, the addition of acids or bases causes only a small change in pH.

cAMP. Second messenger for increased demand for energy and glucose. cAMP activates cAMP-dependent protein kinase. Increased cAMP levels are associated with increased protein phosphorylation. Increases in the cAMP concentration cause activation of glycogen degradation, increased fatty acid breakdown, stimulation of glycolysis in muscle, and stimulation of gluconeogenesis in the liver.

Capping. Putting a 7-methyl-G on the 5′ end of an mRNA molecule. Capping helps the ribosome recognize mRNA and may increase the stability of the RNA.

Carbohydrate. $C_n(H_2O)_n$. Sugar. Carbohydrates have the general formula of a hydrate of carbon, but this doesn't tell you much about the actual structure. All carbohydrates contain one carbon at the aldehyde or ketone oxidation state. This carbon is bonded to another hydroxyl group (alcohol) in the molecule to form a cyclic structure that has a ring size of 5 or 6. The ring size formed depends on the relationship between where the aldehyde or ketone is in the molecule and the total number of carbon atoms. Glucose, which has six carbons with the aldehyde at C-1, forms a six-membered ring. Fructose, which has six carbons with a ketone at C-2, forms a five-membered ring. Ribose, which has five carbons with an aldehyde at C-1, forms a five-membered ring.

γ-Carboxyglutamate–Ca^{2+} binding. A posttranslational modification of proteins, particularly those involved in blood clotting, that involves the carboxylation of certain glutamate residues of the protein in a vitamin K–dependent reaction. γ-Carboxyglutamate formation allows the proteins to bind Ca^{2+}. Not all Ca^{2+}-binding proteins have this modification.

Carboxylate. $R-C(=O)-O^-$. The base formed by dissociation of a carboxylic acid.

Carnitine shuttle. Gets fatty acyl groups into mitochondria. Fatty acyl-CoA in the cytosol is transferred to carnitine to make fatty acyl carnitine, which is transported into mitochondria. Once inside, the fatty acyl group is transferred to CoA and the carnitine is returned to the cytosol. The shuttle is necessary because CoA can't get across the mitochondrial membrane. The shuttle is inhibited by malonyl-CoA, which serves to divert carbon toward fatty acid synthesis and away from oxidation.

Catabolism. Degradative metabolism. Catabolic pathways are turned on by demands for energy.

Catalysis. Increased rate produced by a cyclic reaction process. A catalyst increases the rate of a chemical reaction by using specific interactions between the catalyst and the reactants to stabilize the transition state for the reaction. The catalytic cycle does not consume the catalyst.

cDNA. DNA made from RNA. cDNA is obtained from RNA by using the viral enzyme reverse transcriptase. This enzyme copies RNA templates into DNA–RNA hybrids. After the RNA in these hybrids is specifically destroyed, double-stranded DNA may be produced by DNA polymerase. Since cDNA is a copy of an mRNA, it contains only the exon sequences and usually a poly(A) tail.

Chiral. Asymmetric. Chiral compounds contain asymmetric carbons (or other atoms). A chiral molecule cannot be superimposed upon its mirror image. Carbons (or other elements) having four different sub-stituents on the same carbon are chiral.

Cis. On the same side. Configuration of a double bond in which the two substituents of the double bond are found on the same side of the molecule. This word is also being used to describe the regulatory interactions between two DNA sequences on the same gene. An enhancer or repressor sequence in the DNA is a cis-acting element or factor that affects the transcription of the gene.

Cloning. Copying DNA. Isolating and making exact copies of some piece of DNA you want. A great help in the whole procedure is that you can isolate and then grow a bacterium (or cell) from a single cell. The trick is simply to dilute a population of cells so that there are just a few (say 100) placed on an agar plate or culture dish. Each cell then forms a new colony of cells from which new cultures and billions and billions of cells can be grown—with your DNA going along for the ride.

Cofactor or coenzyme. A molecule bound to an enzyme to help with the chemistry. Most cofactors are vitamins. The cofactors allow the enzyme to have functional groups that are not available from the side chains of the amino acids.

Competitive. Substrate and inhibitor combine at the same site. Since the substrate and inhibitor combine at the same site, a competitive inhib-itor can be completely displaced by a high concentration of the substrate.

Complementation. Using one gene (often present on a vector) to provide a gene product (protein) that is missing from a mutant cell. This pro-cedure can be used to identify the functions of the protein products of specific genes.

Configuration. Used to refer to the stereochemical arrangement of atoms in a molecule. Stereoisomers differ in configuration. Epimers and enantiomers differ in configuration, as do cis and trans isomers around double bonds. Configuration cannot be changed without breaking and reforming covalent bonds. This word is often (incor-rectly) used instead of *conformation*.

Conformation. Differences in rotation around bonds. The spatial arrange-ments of atoms in a molecule that are determined by rotation around covalent bonds. The conformation of a molecule can be changed by simply rotating groups around single bonds. This word is often (in-correctly) used instead of *configuration*.

Cooperativity. The reaction of one substrate affects the reaction of the next substrate with the enzyme or protein. In positive cooperativity, reac-tion of the first substrate makes the reaction of the next substrate easier. In negative cooperativity, the reaction of the first substrate makes the reaction of the next substrate harder. Cooperative effects are mediated by conformation changes in the protein that are brought about when substrate binds to the enzyme.

Cori cycle. Lactate from muscle to liver, glucose from liver to muscle. An interorgan cycle in which lactate made by the muscle is taken up by the liver and converted by gluconeogenesis to glucose. The glucose made by the liver is then recycled to the muscle (and other tissues). The effect is to conserve the glycogen supplies of muscle and to maintain blood glucose levels in response to increased demand by the muscle.

Covalent catalysis. Enzyme forms a covalent bond with the substrate. The amino acid side chains and enzyme cofactors provide functional groups that are used to make the reaction go faster by providing new pathways or by making existing pathways faster. The covalent participation of an enzyme in catalysis requires that the intermediate that is formed be more reactive than the substrate.

Cyanide. Cytochrome oxidase inhibitor. Inhibits electron transport at the last stage, the transfer of electrons to oxygen. It stops oxygen consumption, substrate oxidation, and ATP synthesis.

D. Designation of configuration. For tetrahedral carbon atoms with four different substituents, exchanging any two of them results in an isomer that looks the same chemically but cannot be superimposed on the original structure. The two isomers are called *enantiomers* and are said to differ in configuration. For sugars, the designation of configuration is based on the configuration at C-5 for hexoses and C-4 for pentoses (next to the last carbon). D- and L-glucose differ in configuration at every carbon atom. All D sugars have the OH group on the next to the last carbon pointing to the right when C-1 is on top and C-6 is on bottom as shown in the diagram below. If the OH and H are swapped, the configuration is L.

$$
\begin{array}{c}
\text{C-1} \\
| \\
\text{H}\!-\!\!-\!\text{C}\!-\!\!-\!\text{OH} \\
| \\
\text{C-6}
\end{array}
$$

D. Dextrarotatory. A compound that rotates the plane of polarized light to the right is labeled *d* or (+). The enantiomer of the compound rotates the plane of polarized light to the left by the same amount.

Decarboxylation. Lose CO_2. The reaction catalyzed by a decarboxylase. Decarboxylation of amino acids to give the amines is a frequent way to metabolize them. These decarboxylases invariably use a pyridoxal phosphate cofactor.

Dehydrogenase. Removes electrons and hydrogen from substrate. A reaction in which the named substrate is oxidized (usually by an NAD^+ or FAD) by the removal of a proton and two electrons. Lactate dehydrogenase removes two electrons from lactate (to make NADH). Sometimes this kind of reaction is named in the reverse direction. For example, lactate dehydrogenase could have been named pyruvate reductase, but it wasn't.

Denature. Destroy the secondary, tertiary, and quaternary structure of a protein, DNA, or RNA molecule. To denature a protein, DNA, or RNA, you simply have to change the conditions so that the native, or correctly folded, form is less stable than the denatured form. This may involve changing the temperature, pH, or salt concentration, or it may involve the addition of denaturants such as urea, guanidine hydrochloride, or detergents.

Desaturase. Puts double bonds into fatty acids. Mammals can't put in double bonds farther away than nine carbons from the carboxyl group.

DHFR. Dihydrofolate reductase. Catalyzes the NADPH-dependent reduction of dihydrofolate to tetrahydrofolate. Tetrahydrofolate is needed to transfer one-carbon units during purine and pyrimidine biosynthesis.

Diastereomer. Isomer with a different configuration at one or more (but not all) asymmetric centers in a molecule. If there is more than one chiral center in a molecule, the molecule has stereoisomers that differ in just the configuration about the chiral centers. A molecule with four chiral centers has 2^4 stereoisomers. Diastereomers differ in configuration at one or more carbons but not at all carbons. Enantiomers are mirror images that differ in the configuration at all carbons. Epimers differ in configuration at only one atom.

Dielectric constant. A property of matter that tells you how easy it is to separate charge in a particular medium. Solvents with a high dielectric constant, like water, make it easy to separate two oppositely charged particles. A low dielectric constant (as in the interior of proteins) makes it difficult to separate charge.

Dihedral angle. Angle specifying rotation around a bond.

Disulfide. R–S–S–R. Formed by the oxidation of thiols. Disulfide bonds can be used to covalently link two cysteine residues in a protein. This disulfide cross-link helps stabilize the protein against denaturation.

DNA polymerase. Adds nucleotide to 3′ end of growing chain during DNA synthesis. DNA polymerase requires a primer with a free 3′ end and a template. The new chain is produced by adding the nucleotide dictated by the template to the 3′ end of the growing strand. Synthesis proceeds in the 5′ to 3′ direction as the template strand is read in the 3′ to 5′ direction.

Driving force. The source of the majority of the negative ΔG that makes a chemical reaction occur.

Drug-resistance marker. A gene included on a plasmid or other vector that will produce an enzyme that degrades or detoxifies a drug that can be included in the culture medium. For example, the ampicillin-resistance marker encodes an enzyme (β-lactamase) that hydrolyzes the antibiotic ampicillin. Bacteria that contain the plasmid can be selected by growing them in a medium containing ampicillin. The bacteria that do not have the plasmid do not make the β-lactamase and are killed by the ampicillin.

Eadie-Hofstee. $v = V_{\max} + -K_m(v/[S])$. A linear transformation of the Michaelis-Menten equation. A plot of v against $v/[S]$ has a slope of $-K_m$ and an intercept on the y axis of $V_{\max}$. The x intercept is $V_{\max}/K_m$.

Electrostatic interaction. Interaction between charged groups. The high dielectric constant of water shields electrostatic interactions when they're exposed to water; however, when charged groups are buried, out of contact with water, they are almost always paired with a nearby group of opposite charge, creating a salt bridge.

Elimination. Forms a double bond. Loss of a proton and a group with an electron pair to form a double bond. The conversion of malate [$CH_3CH(OH)CH_2CO_2^-$] to fumarate ($CH_3CH=CHCO_2^-$) is an elimination reaction.

Enantiomer. One of a pair of mirror images. Stereoisomers that differ in configuration at all asymmetric centers. Enantiomers are mirror images. D-glucose and L-glucose are enantiomers. *See D, L, R, S.*

Endergonic. $\Delta G > 0$. The free energy of the products is higher than the free energy of the reactants. Overall, a reaction with a positive ΔG won't happen in the direction written but will go in the opposite direction.

Endothermic. $\Delta H > 0$. A reaction that takes up heat as it occurs. In this case, the enthalpy of products is higher than the enthalpy of the reactants. A reaction with $\Delta H > 0$ cannot happen in the direction written unless the entropy change is favorable enough to make the overall ΔG negative.

-ene. Has a C=C. An ending attached to a chemical name to indicate the presence of one or more double bonds. Pronounced ''een,'' not ''eney.'' A triene has three double bonds.

Enthalpy. ΔH. Heat change during a chemical reaction due to differences in bond energy between products and reactants. Reactions that give off heat ($\Delta H < 0$) make a favorable contribution to ΔG.

Entropy. ΔS. **Disorder.** *Entropy* is just another word for disorder. Any increased order that accompanies a chemical reaction is unfavorable. Disorder is the favored state. An increase in entropy is an increase in disorder, produces a $\Delta S > 0$, and contributes favorably to the overall change in free energy.

Epimer. Stereoisomer differing in the configuration about one asymmetric center. For compounds that have multiple asymmetric centers, stereoisomers that differ in the configuration at just one of the centers (the rest are the same) are called *epimers*. Galactose is a C-4 epimer of glucose; galactose and glucose differ only by the configuration of the hydroxyl group at C-4.

Epitope. The feature of an antigen that is recognized by the active site of an antibody molecule.

Epimerase. Changes the configuration. Epimerases change the configuration of the substrate at a single carbon. To convert glucose to galactose you only have to exchange the hydrogen and OH groups at C-4. This is an epimerization catalyzed by an epimerase (UDP-glucose epimerase).

Epinephrine. Adrenalin. Excitement hormone. Indicates the need for immediate energy. Turns on glycogenolysis and glycolysis in muscle, turns on glycogenolysis and gluconeogenesis in liver, and turns on lipolysis in adipose tissue.

Equilibrium constant. $K_{eq} = [\text{products}]_{eq}/[\text{reactants}]_{eq}$. Ratio of the concentration of products to reactants at chemical equilibrium. The bigger the equilibrium constant, the more intrinsically favorable the reaction.

Ester. RC(=O)-OR. A functional group in which an acyl group is attached to an alcohol.

Esterase. Hydrolyzes esters. An enzyme that hydrolyzes esters to an alcohol and a carboxylic acid.

Exergonic. $\Delta G < 0$. The free energy of the products is lower than the free energy of the reactants. Overall, this reaction will happen in the direction written.

Exothermic. $\Delta H < 0$. Reaction that gives off heat as it occurs. The enthalpy of the products is lower than the enthalpy of the reactants. Overall this reaction will happen in the direction written unless the entropy change is unfavorable enough to make the ΔG positive.

FAD. Flavin adenine dinucleotide; $FADH_2$ is the reduced form. A cofactor for oxidation and reduction reactions. Recognize it by three fused rings with a lot of nitrogens in them. Succinate dehydrogenase is an FAD-linked enzyme. Oxidation of $FADH_2$ by the electron transport chain is energetic enough to make two ATPs.

Fasting. Not eating. During a short, overnight fast, glycogen provides glucose equivalents and fat provides energy. The glucagon level is high; the insulin level is low. Protein phosphorylation is increased.

Fat. Energy-storage form. Mainly stored as triglyceride in adipose tissue. When energy is needed, the adipose tissue releases fatty acids by the activation of a hormone-sensitive lipase (stimulated by glucagon and epinephrine) that catalyzes the hydrolysis of the triglyceride. The fatty acids are then transported through the serum and oxidized via β oxidation in the tissues to yield energy.

Fatty acid synthase. Multifunctional enzyme for fatty acid synthesis. Catalyzes the synthesis of C_{16} fatty acid (palmitate) from acetyl- and malonyl-CoA. The net reaction is acetyl-CoA + 7 malonyl-CoA + 14NADPH + 14H$^+$ → C_{16} fatty acid + 14NADP$^+$ + 8 CoA. Fatty acid synthesis overall is regulated by the activity of acetyl-CoA carboxylase.

FBPase. Fructose 1,6-bisphosphatase. This regulated enzyme of gluconeogenesis catalyzes the hydrolysis of fructose 1,6-bisphosphate to fructose 6-phosphate. Inhibited by fructose 2,6-bisphosphate (a signal for high glucose levels).

First order. A → B. $v = k[A]$ (k in s^{-1}). In first-order reactions, the velocity is proportional to the concentration of substrate present at any time. Doubling the substrate concentration doubles the rate. The actual concentration of substrate at any time during the reaction is given by $A = A_0 e^{-kt}$.

Fischer projection. A two-dimensional representation of the tetrahedral geometry of carbon. In a Fischer projection, the bonds to carbon are represented in a two-dimensional form. The bonds represented as vertical are actually behind the plane of the paper, and the bonds shown as horizontal are in front of the plane of the paper.

Flavin. A coenzyme involved in oxidation-reduction. Comes in several flavors: riboflavin, flavin adenine dinucleotide (FAD), and flavin mononucleotide (FMN). The name depends on what's attached to the nonfunctional part of the flavin. Flavins can transfer electrons one at a time or two at a time, which makes them ideally suited to couple electron transfer from NADH to the one-electron carriers of the electron transport chain.

Free energy. $\Delta G = \Delta H - T\Delta S$. Useful energy that can be derived from a chemical reaction. ΔG is the energy that drives a chemical reaction. A reaction is favorable (proceeds in the direction written) when ΔG is negative (< 0). It's a combination of two factors—the change in chemical bond energy (ΔH) that occurs during the reaction and the change in disorder (ΔS). Reactions that give off heat ($\Delta H < 0$) and result in disorder ($\Delta S > 0$) are more favorable. If ΔG is positive, the reaction goes in a direction opposite the one written, and if ΔG is zero, the reaction is at chemical equilibrium.

$$\Delta G = -RT \ln K_{eq} + RT \ln \frac{[\text{products}]}{[\text{reactants}]}$$

Furanose. A carbohydrate in a cyclic five-membered ring. Formed by sugars that have five or six carbon atoms. For sugars with six carbons, a furanose forms when the carbonyl is at C-2 (as in fructose).

Fructose 1,6-bisphosphatase. FBPase. This regulated enzyme of gluconeogenesis catalyzes the hydrolysis of FBP to fructose 6-phosphate during gluconeogenesis. Inhibited by fructose 2,6-bisphosphate.

Fructose 2,6-bisphosphate. Alternative signal for glucose. A regulator of glycolysis-gluconeogenesis that signals a high glucose level. Stimulates glycolysis (phosphofructokinase) and inhibits gluconeogenesis (fructose 1,6-bisphosphatase). Made from fructose 6-phosphate and ATP by phosphofructo-2-kinase.

Futile cycle. Hydrolyzes ATP. A group of enzyme reactions that would hydrolyze ATP if all the enzymes were active at the same time. Phosphofructokinase and fructose 1,6-bisphosphatase would constitute a futile cycle if both enzymes were active at the same time.

$$\text{F-6-P} + \text{ATP} \longrightarrow \text{F-1,6-P}_2 + \text{ADP}$$
$$\underline{\text{F-1,6-P}_2 + \text{H}_2\text{O} \longrightarrow \text{F-6-P} + \text{P}_i}$$
$$\text{Net:} \quad \text{ATP} + \text{H}_2\text{O} \longrightarrow \text{ADP} + \text{P}_i$$

Futile cycles may be used to generate heat by ATP hydrolysis or they may be used to enhance the sensitivity of the reactions to regulation. Since you can affect the flow through the reaction by turning up one enzyme while turning down the opposing one at the same time, these cycles can be made very responsive to regulatory signals.

Gel. A way to separate protein, DNA, or RNA molecules based on their size and/or charge. Usually a flat slab of gel (the gel looks much like tough gelatin dessert) is poured between two glass plates, the sample is applied at the top, and an electric charge is applied across the gel to cause the charged molecules to move. The gel network retards big molecules more than small ones so that small molecules move faster.

Genomic DNA. Fragments of DNA from the genome of some organism. This is DNA that has been isolated directly from nuclear DNA of some organism. It contains exons, introns, 5′-untranslated regions—anything that can occur in DNA.

Glucagon. Low-glucose signal. Increases cAMP level. A peptide hormone that signals low blood glucose levels. Through activation of adenylate cyclase, glucagon activates glycogenolysis and gluconeogenesis in the liver and lipolysis in the adipose tissue.

Glucogenic. Amino acids that can be converted to glucose equivalents. Most amino acids (except Leu and Lys) can be metabolized to give something that can be converted to glucose. These amino acids serve as a source of glucose after the body has exhausted its glycogen stores.

Gluconeogenesis. Making glucose from pyruvate. To maintain blood glucose levels, liver and kidney can convert pyruvate to glucose and release it to the rest of the body. Gluconeogenesis is activated by low blood glucose levels in response to glucagon.

Glucose. $HC(=O)-CH(OH)-CH(OH)-CH(OH)-CH(OH)-CH_2(OH)$. Input substrate for glycolysis, output product of gluconeogenesis. Glucose (its metabolites) is required to replenish intermediates of the TCA cycle and as an energy source for red cells and brain.

Glucose 6-phosphatase. Required for gluconeogenesis. This enzyme hydrolyzes glucose 6-phosphate to glucose as the last step of gluconeogenesis. The enzyme is missing from muscle, so muscle cannot make glucose.

Glycogen. Glucose storage. A branched polymer of glucose in which the glucose residues are linked to each other in a 1-4 or a 1-6 linkage (branch points). Provides a 24-hour reserve of glucose to meet the demands for glucose during fasting.

Glycogen branching enzyme. Creates a branch point (1-6 linkage) in glycogen. Branching enzyme actually accomplishes this the hard way, by moving multiple glucose residues from a 1-4 branch to the 6-position of a specific glucose residue.

Glycogen synthase. $(Glycogen)_n + UDP\text{-}glucose \rightarrow (Glycogen)_{n+1} + UDP$. Enzyme adds a glucose unit in a 1-4 linkage to the end of a glycogen chain. The enzyme is inactivated by phosphorylation; however, the unphosphorylated enzyme can be activated to make glycogen by high concentrations of glucose 6-phosphate.

Glycolysis. $Glucose + 2ADP + 2P_i + 2NAD^+ \rightarrow 2\ pyruvate + 2ATP + 2NADH + 2H^+$. Metabolic pathway that provides pyruvate as fuel to the TCA cycle or for fat synthesis. Complete metabolism to CO_2 yields 38 ATPs per glucose (assuming 3 ATPs for each cytosolic NADH). In the absence of oxygen, lactate is produced from the pyruvate to regenerate NAD^+ so that the pathway can continue to work in the absence of oxygen, generating only 2 ATPs per glucose. Glycolysis is turned on by low-energy or high-glucose signals. Fructose 2,6-bisphosphate activates glycolysis by activating phosphofructo-1-kinase.

Glucose 6-phosphate dehydrogenase. Entry into the hexosemonophosphate

pathway. **Glucose 6-phosphate + NADP$^+$ → 6-phosphogluconolactone + NADPH + H$^+$.** Catalyzes the conversion of glucose 6-phosphate to 6-phosphogluconolactone, making an NADPH in the process. Inhibited by NADPH, one of the products of the HMP pathway.

Glycoprotein. Protein with carbohydrate attached. There are two ways to attach branched carbohydrate polymers to proteins. If the carbohydrate is attached to an Asn residue (through the nitrogen), it is called N-linked. N-linked carbohydrate is initially added as a preformed block of polysaccharide using dolichol pyrophosphate as a polysaccharide carrier and donor. Carbohydrate can also be attached to Ser or Thr residues. These are called O-linked. O-linked carbohydrate is built up by adding one sugar residue at a time. The carbohydrate serves a number of roles including recognition and cell–cell communication.

Guanidine. H$_2$N–C($=$NH$_2^+$)–NH$_2$. At neutral pH, guanidine exists as the protonated form, the guanidinium ion. At high concentrations (2–6 M), guanidinium hydrochloride increases the solubility of the hydrophobic side chains of the amino acids and causes proteins to lose their structure and denature.

Half-life: The time it takes half of something to go away.

Hanes-Wolf. [S]/v = K_m/V_{max} + [S]/V_{max}. [S]/v vs. [S]. A transformation of the Michaelis-Menten equation that gives a straight line plot of [S]/v against [S].

Helicase. Unwinds DNA. An enzyme activity involved in DNA replication that relieves the strain (supercoiling) associated with unwinding the DNA double helix during replication.

Heme. A cofactor consisting of a porphyrin ring containing an iron atom. The porphyrin ring provides four nitrogen ligands for the iron atom. The other two ligands are provided by the protein or the substrates. The hemes have different functions depending on the protein that uses them as a cofactor. Hemes are used to carry oxygen without oxidizing it in hemoglobin and myoglobin, but in other proteins, like cytochrome P450, the heme iron produces a very reactive iron–oxygen species at the active site.

Hemithioacetal. Product from the addition of a thiol (RSH) to an aldehyde. RS–CH(OH)–R. A hemithioacetal is formed between the enzyme and the substrate, glyceraldehyde 3-phosphate, during the reaction catalyzed by glyceraldehyde-3-phosphate dehydrogenase.

Henderson-Hasselbalch equation. pH = pK_a + log ([base]/[acid]). Describes the behavior of acids and bases including titration and buffering. If the pH is bigger than the pK_a, there must be more of the basic form of the buffer present in solution than the acidic form.

Hexokinase. Glucose + ATP → glucose-6-P + ADP. Responsible for the phosphorylation of glucose for entry into glycolysis, glycogen synthesis, or the pentose phosphate pathway. Hexokinase is inhibited by its product, glucose 6-phosphate.

Hexose. A six-carbon sugar. Glucose and fructose are hexoses.

HMP pathway. The hexose monophosphate pathway converts glucose 6-phosphate to C-3, C-4, C-5, and C-7 sugars and produces NADPH for biosynthetic reduction needs. Inhibited by NADPH.

hnRNA. Heterogeneous nuclear RNA. RNA formed in the nucleus that is a precursor to processed forms of RNA (e.g., mRNA). Also called the *initial transcript*. hnRNA has both the intron and exon sequences and poly(A) addition sites.

Hormones. Signal molecules of the endocrine system.

Hydrogen bond. Sharing a hydrogen atom between one atom that has a hydrogen atom (donor) and another atom that has a lone pair of electrons to accept the hydrogen bond (acceptor).

Hydrophobic interaction. Hydrophobic groups cannot participate in the hydrogen-bonding network of water. In order to minimize the extent of contact between hydrophobic groups and water, hydrophobic groups aggregate with each other. This minimizes the number of water molecules that are in contact with a hydrophobic surface. When hydrophobic groups associate, the water molecules that had restricted freedom at the contact surface are free to be normal water molecules. This increase in water entropy makes the association between hydrophobic molecules favorable.

3-hydroxy-3-methylglutaryl-CoA reductase. HMG-CoA + 2NADPH + $2H^+$ → mevalonate + $2NADP^+$. Regulated enzyme of isoprenoid biosynthesis including cholesterol. Regulated by phosphorylation (inactivated) and mRNA levels.

Hydroxylysine. A posttranslational modification of lysine that occurs in collagen but not in elastin. The hydroxyl groups are added by lysyl hydroxylase and serve as the point of attachment of O-linked polysaccharides. Like lysine, they also participate in the cross-linking of collagen.

Hydroxyproline. A posttranslational modification of proline that occurs in collagen and to a smaller extent in elastin. The hydroxyl groups are added by prolyl hydroxylase and serve as the point of attachment of O-linked polysaccharides.

Hyperbolic kinetics. $v = V_{max} [S]/(K_m + [S])$. A description of the dependence of the velocity of an enzyme-catalyzed reaction on the concentration of the substrate. If an enzyme follows hyperbolic kinetics, it is described by the Michaelis-Menten equation.

β-hydroxybutyrate. CH$_3$CH(OH)CH$_2$CO$_2^-$. A ketone body. Formed by the liver to free up CoA for more β oxidation. Used by peripheral tissues after conversion to acetoacetyl-CoA.

Imidazole. Histidine side chain. Imidazole is the aromatic, nitrogen-containing side chain of histidine. It can be protonated (on either nitrogen). The pK_a is 6 to 7. This side chain is often used as a general acid or general base catalyst by enzymes to help remove or donate protons during the chemical reaction.

Immunoprecipitation. Using antibodies to precipitate a specific protein. Binding of specific parts of a protein (epitopes) to the active sites of a collection of antibodies leads to a network of cross-linked protein that becomes insoluble and precipitates from solution. To actually precipitate, the antibody preparation must contain different antibodies that recognize multiple epitopes of the antigen. Monoclonal antibodies that recognize only one region of a protein aren't normally able to immunoprecipitate the protein; however, it's possible to use monoclonal antibodies to pull an antigen out of solution by attaching the monoclonal to an insoluble support.

Induced fit. Protein changes conformation. A model to explain the specificity of enzymes in which a good substrate has enough binding interaction with the enzyme to induce a conformation change that brings the catalytic groups into the right position for efficient function. Bad substrates don't have an extensive enough binding interaction with the enzyme to cause the conformation change.

Inhibitors. Molecules that resemble the substrate(s) and bind or receptors to the enzyme active site. Inhibitors fill up the active site and prevent the normal interaction with substrate. Reversible inhibitors simply bind using hydrophobic interaction, hydrogen bonds, et cetera. Irreversible inhibitors react chemically with the enzyme and inactivate it permanently (until more of the protein can be synthesized).

Inhibitors of oxidative phosphorylation. These block the flow of electrons at a specific site or inhibit the exchange of ATP and ADP from the mitochondria. Inhibitors of oxidative phosphorylation inhibit oxygen consumption, substrate consumption, and the synthesis of ATP.

Initial velocity. The velocity of an enzyme-catalyzed reaction that is measured under conditions where there is no significant change in the concentration of substrate.

Insert. The piece of DNA you have placed (inserted) into a vector using restriction endonucleases.

Insulin. Glucose signal secreted by the pancreas. Insulin activates glucose uptake, glycolysis, fatty acid synthesis, glycogen storage, and protein synthesis. How insulin actually causes these changes is not known. When it binds insulin, the insulin receptor becomes an active protein kinase (tyrosine-specific); however, other substrates for the protein kinase activity have not been directly identified.

Isoelectric point. pH at which a molecule has zero charge. If a molecule has both acidic ($-$ charged) and basic ($+$ charged) groups, there is a pH at which the molecule has no charge. The pI can be approximated by averaging the two pK_a's that interconvert the molecules with $+1$ to 0 charge and 0 to -1 charge. If pH $<$ pI, the molecule is positively charged. If pH $>$ pI, the molecule is negatively charged. (Remember that adding a proton makes the molecule more positively charged.)

Isomerase. An enzyme that catalyzes an intramolecular rearrangement.

Ketogenic amino acids. Lys and Leu. Amino acids that are degraded only to acetyl-CoA or other molecules that cannot be used to synthesize glucose or TCA-cycle intermediates.

Ketone. R–C(=O)–R. Fructose has a ketone (at C-2).

Ketone bodies. Acetoacetate and hydroxybutyrate. Synthesis of ketone bodies frees up CoA for more β oxidation in the liver. Ketone bodies are metabolized in muscle and brain (after adaptation) as an energy source.

Ketose. A sugar that has a ketone [R–C(=O)–R] functional group when it is written in the open-chain form. Fructose is a ketose. *See* aldose.

Kinase. Substrate + ATP → substrate–O–P + ADP. Kinases incorporate phosphate from ATP into the substrate.

K_m. [S] where $v = V_{max}/2$. If an enzyme follows hyperbolic kinetics, the K_m is the concentration of substrate necessary to give a velocity that is one-half V_{max}. K_m has units of M. It can be a true dissociation constant for formation of the ES complex; however, it may not be a true dissociation constant, depending on how big certain rate constants are with respect to each other.

L. Designation of configuration. For compounds with assymetric carbon atoms (carbon atoms with four different substituents), L and D are used to designate the spatial arrangement of the substituents around a given carbon atom (configuration). The system relates the configuration at the assymetric center to the configuration of L- or D-glyceraldehyde.

l. **Levorotatatory.** An optically active compound that rotates plane polarized light to the left. Chiral compounds (compounds that have non-superimposible mirror images) are optically active. There is no correlation between D and *d* or L and *l*. *d* and *l* are just experimental observations. D and L are absolute statements of the structure of the stereoisomer.

Libraries. Collections of DNA sequences in some vector. These are usually a random collection of fragments that represent the DNA or RNA in an organism. There are two kinds of libraries. cDNA libraries are created by isolating a mixture of mRNAs from an organism. The mRNAs are converted to DNA with reverse transcriptase and then inserted into a vector (one DNA piece per vector). Genomic libraries are generated by cutting the total genomic DNA with restriction enzymes and inserting the pieces into some vector. Don't forget that libraries are collections. To get the DNA piece you want out of the library, you must have some screening or selection method that will allow you to identify the cell that contains the piece of DNA you want.

Ligand. A molecule that binds to a receptor or enzyme. In inorganic chemistry, the ligand is the thing that does the binding; however, in biochemistry, the ligand is the thing that is bound. Remember *backward?*

Ligation. Joining two pieces of DNA covalently. The enzyme DNA ligase joins the backbone phosphates in a phosphodiester bond. This is an ATP-requiring reaction.

Lineweaver-Burk. $1/v = 1/V_{max} + (K_m/V_{max})(1/[S])$. A linear transformation of the Michaelis-Menten equation. The slope of a plot of $1/v$ against $1/[S]$ is K_m/V_{max}; the intercept on the $1/v$ axis is $1/V_{max}$. The intercept on the $1/[S]$ axis is $-1/K_m$.

Lipase. A lipase catalyzes the hydrolysis of the ester bond that attaches fatty acids to triglycerides or phospholipids. Hormone-sensitive lipase is an enzyme of the adipocytes that releases fatty acids in response to epinephrine and glucagon signals (low energy, low glucose levels). Lipoprotein lipase is an enzyme of the epithelia that releases fatty acids from lipoproteins so they can be taken up by the tissues for storage or use.

Lipids. Biological molecules soluble in organic solvents like chloroform or ether. Lipids consist of a diverse set of hydrophobic molecules including triglycerides, phospholipids, steroids, and so forth.

Lipoic acid. Cofactor for pyruvate and α-ketoglutarate dehydrogenases. This cofactor contains two sulfurs in a five-membered ring. The cofactor can function both as an oxidation-reduction cofactor (thiol and disulfide forms) and as an acyl carrier [R–C(=O)–S–lipoic acid].

Lock and key. Specificity model. A model developed by Emil Fischer to account for the specificity of enzymes. In this model, the enzyme is complementary to the substrate and recognizes the correct substrate as a lock recognizes a key.

Logarithms. The answers you get on your calculator when you press the log or ln button. The power to which 10 or e (depending on the base of the logarithms) has to be raised to give the number you entered in the calculator. $10^{\log x} = x$. For base-10 logarithms, a difference in 1 log unit means that the actual numbers differ by a factor of 10. Note that $\log_{10} x = (\ln x) [\ln 10)]$, or $\log_{10} x = \ln x/2.303$.

Lyase. Cleaves C—C, C—O, or C—N bonds without hydrolysis or oxidation-reduction. For example, citrate lyase cleaves citrate into oxaloacetate and acetyl-CoA.

Malate-aspartate shuttle. Gets electrons from cytoplasmic NADH into the mitochondria so that 3 ATPs can be made by oxidation of the NADH.

Malic enzyme. Pyruvate + CO_2 + $NADP^+$ → malate + NADPH + H^+. An anaplerotic reaction that replenishes the intermediates of the TCA cycle using pyruvate generated from glucose or other carbohydrates.

Metabolic acidosis. Decreased serum pH, caused by a decrease in the serum bicarbonate concentration.

Metabolic alkalosis. Increased serum pH, caused by an increase in the bicarbonate concentration.

Michaelis-Menten equation. $v = V_{max} [S]/(K_m + [S])$. Describes the dependence of the velocity of an enzyme-catalyzed reaction on the concentration of substrate. At low substrate concentrations, the velocity increases linearly with increasing substrate concentration. At very high substrate concentrations ($[S] \gg K_m$), the velocity approaches V_{max}.

Mixed-function oxidase. An enzyme that catalyzes the incorporation of one oxygen atom (as OH) from molecular oxygen (O_2) into the substrate.

Monoclonal antibody. An antibody that recognizes a single epitope. All antibody molecules in the monoclonal population have identical combining sites. They are produced by expressing the DNA from a single antibody-producing B cell.

mRNA. Messenger RNA. The primary RNA transcript is processed to mRNA by adding a 5′ cap (7-methyl-G) and a poly(A) tail and removing introns (splicing).

Mutase. An enzyme that catalyzes an intramolecular rearrangement.

NAD^+-NADH. Nicotinamide adenine dinucleotide. NADH is an electron carrier. NAD^+ accepts two electrons (and a proton) from substrates and ultimately donates them to the electron transport chain to make three ATPs. NAD^+ is the oxidized form of the cofactor. The + on NAD^+ is used to designate the charge on the pyridine ring of the nicotinamide; the actual molecule is negatively charged.

$NADP^+$-NADPH. Nicotinamide adenine dinucleotide phosphate. NADPH is generated by the HMP pathway. In contrast to NADH, NADPH is used mainly for biosynthetic reductions. The oxidized form of the cofactor is $NADP^+$

Negative nitrogen balance. Excreting more nitrogen from the body than you take in from the diet. Negative nitrogen balance may occur when the diet is missing an essential amino acid. Stored proteins are degraded to supply the missing essential amino acid, and the extra nonessential amino acids that come from the degraded protein are metabolized and the nitrogen excreted.

Negative selection. A selection in which the cells having the DNA of interest die in the selection process and cells without the DNA of interest live. Obviously, this selection must be done on replica plates. Replica plates are made by transferring a few cells from each colony to a new culture plate. The transfer is done by putting a piece of filter paper onto the original dish, removing it, and then placing the filter paper (and cells from each colony) onto a new dish. Since an exact replica is made, the colonies that die under selection on the replica plate can be found on the original plate.

Nicotinamide. A vitamin that serves as a source of the pyridine ring of NAD^+ and $NADP^+$.

Noncompetitive inhibition. A type of enzyme inhibition in which the inhibitor does not prevent the binding of the substrate to the enzyme. On a Lineweaver-Burk plot, the pattern of lines intersect to the left of the $1/v$ axis.

Nonproductive binding. Specificity model. Poor substrates bind to the enzyme in a large number of different ways, only one of which is the proper one for catalysis.

Northerns. RNA detected with DNA probe. RNA is run out on an agarose or acrylamide gel and blotted to nitrocellulose paper, and specific sequences in the DNA are detected using a labeled DNA probe.

Oligomycin. Inhibitor of oxidative phosphorylation that prevents the phosphorylation of ADP.

Oligonucleotide. A small piece of DNA (*oligo* means "just a few" whereas *poly* means "a bunch"). The place where an oligonucleotide becomes a polynucleotide is not rigidly defined. Oligonucleotides (2 to 30 or more nucleotides) can be synthesized chemically using high-yield and specific reactions that are performed automatically on a solid matrix.

Operon. A collection of genes that are clustered together and whose expression is controlled by the same regulatory region of the DNA. This arrangement allows simple control over the expression of proteins that are all needed for a common job.

Optical isomers. Stereoisomers. Optical isomers, or stereoisomers, differ in the configuration around one or more asymmetric centers in the molecule.

Oxidase. An enzyme that catalyzes the incorporation of oxygen into the substrate.

Oxidation. The loss of electrons. NADH is oxidized to NAD^+, Fe^{2+} is oxidized to Fe^{3+}. When something is oxidized, something else must become reduced.

β oxidation. Metabolism of fat. The pathway produces acetyl-CoA from fat. Regulated by *carnitine acyl transferase,* which is *inhibited by malonyl-CoA.* A series of reactions occurring in the mitochondria in which a fatty acyl-CoA is first oxidized to place a double bond at C-2, then hydrated, oxidized to the 3-keto fatty acid, then cleaved by a CoASH-dependent reaction into acetyl-CoA and a fatty acid that is two carbons shorter.

Oxidoreductase. An enzyme that catalyzes an oxidation or a reduction.

P/O ratio. The number of ATP equivalents made per two electrons passed down the electron transport chain. Each NADH or $FADH_2$ oxidized by electron transport counts as one O (two electrons).

Palindrome. Reads the same backward as forward. For DNA, *backward* means on the other strand in the opposite physical direction. It doesn't mean backward on the same strand.

```
5'——————————ATGCAT——————————3'
3'——————————TACGTA——————————5'
```

PCR. Polymerase chain reaction. A method to amplify a specific piece of DNA between two oppositely oriented primers.

Pentose. A five-carbon sugar. Ribose is a pentose.

Peptidase. Hydrolyzes the peptide bond in peptides and proteins.

PFK. Phosphofructokinase. Fructose 6-phosphate + ATP → Fructose 1,6-bisphosphate + ADP. Enzyme of glycolysis turned on by high-glucose signals ($F-2,6-P_2$) and low-energy signals. *See* phosphofructokinase.

pH = $-\log[H^+]$. A measure of the concentration of the strongest acid in water, H_3O^+ (abbreviated H^+). The lower the pH, the higher the concentration of H^+. A change in pH of 1 unit corresponds to a 10-fold change in $[H^+]$.

Phenol. An OH group attached to a benzene ring. The amino acid tyrosine is a phenol.

Phenotype. A behavior or characteristic of an organism that is detectable.

Phosphatase. Hydrolyzes a phosphate ester. $R-OPO_3H^- + H_2O \rightarrow R-OH + P_i$.

Phosphate diester. A phosphate with two alcohols attached to the phosphorus. $RO-PO_2-OR^-$. DNA is a phosphate diester. Also referred to as a *phosphodiester*.

Phosphate monoester. A phosphate with one alcohol attached to the phosphorus. $RO-PO_3H^{2-}$. Sugar phosphates such as glucose 6-phosphate are phosphate monoesters. Also referred to as a *phosphomonoester*.

Phosphate triester. A phosphate with three alcohols attached to the phosphorus. $RO-PO(OR)-OR$. Phosphate triesters are rare in biology.

Phosphodiesterase. An enzyme that hydrolyzes phosphodiesters. DNA and RNA endo- and exonucleases are phosphodiesterases.

Phosphoenol pyruvate carboxykinase. Oxaloacetate + GTP → PEP + CO_2 + GDP. Regulated enzyme of gluconeogenesis.

Phosphofructokinase. PFK. Fructose 6-phosphate + ATP → Fructose 1,6-bisphosphate + ADP. Unfortunately there are two PFKs. PFK-1 catalyzes the reaction that controls glycolysis, the formation of fructose 1,6-bisphosphate. PFK-2 catalyzes the formation of the regulator of glycolysis, fructose 2,6-bisphosphate. PFK-1 and glycolysis are activated by *low-energy* and *high-glucose* signals. Fructose 2,6-bisphosphate (a signal for high glucose) activates PFK-1.

Phosphorylase. (Glycogen)$_n$ + P$_i$ → glucose 1-phosphate + (glycogen)$_{n-1}$. Rate-limiting enzyme of glycogenolysis. Phosphorylase is activated in response to both low-glucose and low-energy signals. *Phosphorylation activates* the enzyme.

Phosphorylation. Making a phosphomonoester. ROH + ATP→ $RO-PO_3H^-$ + ADP. The reaction can happen to sugars during glycolysis and to proteins (on Ser, Thr, or Tyr) as a regulatory mechanism.

p*I*. Isoelectric point. pH at which molecule has no net charge. For molecules with multiple ionizable acidic (− charged) and basic (+ charged) groups, there is some pH at which the positive and negative charges on the molecule are equal. This is the p*I*. The p*I* can be estimated by averaging the pK_a that converts the molecule with + 1 charge to 0 charge with the pK_a that converts 0 charge to − 1 charge.

pK_a. A measure of the strength of an acid. $-\log(K_a)$. K_a is the equilibrium constant for the acid dissociation reaction (HA $\rightleftharpoons$ H$^+$ + A$^-$). K_a = [H$^+$][A$^-$]/[HA]. The pK_a is calculated from the K_a by taking the negative logarithm. This means that the stronger the acid, the lower the pK_a.

PKU. Phenylketonuria. A genetic defect in the enzyme phenylalanine hydroxylase or the enzyme dihydrobiopterin reductase. Characterized by a high level of phenylpyruvate in the urine resulting from the inability to metabolize phenylalanine. Tyrosine is an essential amino acid in these folks.

Plasmids. Autonomously replicating circular pieces of DNA, which often carry drug-resistance markers.

Poly(A) tailing. The addition of a stretch (50 to 200) of A residues at the 3′ end of an mRNA. Poly(A) tailing requires a signal in the DNA that directs the cutting of the DNA and a poly(A) polymerase that adds the A residues without needing a template.

Polyclonal antibody. A collection of antibodies from the serum of an animal that has been exposed to an antigen. For most antigens, the antibodies in the serum recognize multiple sites (epitopes) on the antigen.

Polysaccharides. Polymers of sugars.

Positive cooperativity. The binding or reaction of the first substrate molecule with an enzyme or receptor makes the reaction of the next substrate easier.

Positive selection. A selection in which the cells having the DNA of interest survive the selection. This is often achieved by including a gene that codes for drug resistance and growing the cells in a culture that contains the drug. Cells that have incorporated the drug-resistance gene survive while those that haven't don't.

Promoter. A region of the DNA to which RNA polymerase binds to begin transcription. The promoter is on the 5′ side of the transcription start site.

Protease or proteinase. An enzyme that hydrolyzes the amide bonds in a protein. Most proteases recognize a specific type of amino acid side chain and cleave the protein at specific points. Serine proteases use an active-site serine to catalyze cleavage, thiol proteases use a cysteine, aspartic proteases use two aspartates, and metalloproteases use a metal such as Zn^{2+} to catalyze the reaction.

Protein kinase. An enzyme that catalyzes the phosphorylation of a Ser, Thr, or Tyr hydroxyl group in a protein. This modification usually affects the function of the protein. Not all Ser, Thr, or Tyr residues in a protein can be phosphorylated. It depends on the sequence surrounding the Ser, Thr, or Tyr and whether or not the protein kinase can get access to the specific hydroxyl group.

Protein. A linear polymer of amino acids connected by amide bonds.

Pyranose. A carbohydrate in a cyclic six-membered ring. Formed only with sugars that have six or more carbon atoms.

Pyridoxal phosphate. A coenzyme that is involved in almost any reaction that involves an amino acid—transamination, decarboxylation, racemization, and elimination.

Pyruvate carboxylase. Pyruvate + CO_2 + ATP → oxaloacetate + ADP + P_i. An anaplerotic reaction that replaces the intermediates in the TCA cycle. A biotin-dependent carboxylase.

Pyruvate dehydrogenase. Pyruvate + NAD^+ + CoA → acetyl-CoA + NADH + H^+. The main connection between glycolysis, the TCA cycle, and fatty acid synthesis. It's a multienzyme complex that contains thiamin, lipoic acid, and flavin as cofactors. The complex also comes with a built-in kinase and phosphatase that control the activity. If there's enough acetyl-CoA around, the enzyme is inactivated (by phosphorylation by the kinase). The dehydrogenase is activated by NAD^+, ADP, and CoA.

Pyruvate kinase. PEP + ADP → pyruvate + ATP. One of the ATP-generating reactions of glycolysis. The reaction is essentially irreversible. Gluconeogenesis must then go by another pathway, not the simple reverse of glycolysis.

***R*. Designation of configuration.** A systematic nomenclature system for specifying the absolute stereochemical configuration at an asymmetric center. Serves the same purpose as the DL system. The other enantiomer would be designated as *S*. Used as a prefix, as in (*R*)-lactate.

Rate constant. A proportionality constant that allows you to calculate the velocity of a chemical reaction from the concentration of substrates. The faster the reaction, the bigger the rate constant. First-order rate constants have units of $time^{-1}$. Second-order rate constants have units of $M^{-1} time^{-1}$.

Recombination. Information swapping by breaking and joining chromosomal DNA. Recombination can occur between genes with similar sequences (*homologous*) or between genes with different sequences (*nonhomologous*). The site of recombination can occur at the exact same spots on two identical chromosomes (*aligned*) or at different spots, even on different chromosomes (*nonaligned*). Nonhomologous or nonaligned recombination results in the duplication or deletion of genetic material in the daughter chromosomes.

Reductase. An enzyme that catalyzes a reduction of the named substrate. Dihydrofolate reductase catalyzes the reduction of dihydrofolate to tetrahydrofolate.

Reduction. The gain of electrons. O_2 is reduced to H_2O. NAD^+ is reduced to NADH. Fe^{3+} is reduced to Fe^{2+}.

Replication. Generation of a new copy of double-stranded DNA from a parental DNA molecule. Replication begins at unique sites (origins), proceeds bidirectionally, and is semiconservative (each daughter molecule has one old strand and one newly synthesized strand). DNA synthesis during replication occurs in the 5' to 3' direction.

Respiratory acidosis. pCO_2 is high. An acid–base imbalance in which the concentration of CO_2 in the blood is high. Caused by depressed respiration.

Respiratory alkalosis. pCO_2 is low. An acid–base imbalance in which the concentration of CO_2 in the blood is low. Caused by rapid and deep respiration.

Restriction endonuclease. Enzyme that cuts DNA at specific recognition sites (restriction sites). Cleavage sequences are usually specified by writing only one strand and showing the cleavage site with an arrow, but both strands are cut. If you write the sequence of the complementary strand, you'll find that if you read the bottom strand sequence in the 5′ to 3′ direction it will be identical to the sequence of the top strand. The cut on the lower strand occurs between the same two nucleotides as in the top strand. The first few letters of the enzyme name denote the organism from which the enzyme is isolated. The next few letters and roman numerals provide a way to tell different enzymes from the same organisms apart.

Retrovirus. A virus containing an RNA genome.

Reverse transcriptase. Makes DNA using an RNA template.

Ribose. A five-carbon sugar. Used in RNA or, in its 2′-deoxy version, in DNA. Made by the hexose monophosphate (HMP) pathway.

Rotenone. Inhibitor of oxidative phosphorylation that blocks NADH dehydrogenase and prevents the oxidation of NADH and the substrates linked to it. However, substrates like succinate that enter via $FADH_2$ are oxidized and make two ATPs per mole.

S. Designation of configuration. A systematic nomenclature system for specifying the absolute stereochemical configuration at an asymmetric center that serves the same function of the DL system. The other enantiomer would be R. Used as a prefix, as in (S)-lactate. *See R.*

β sheet. Secondary structure. Element of secondary structure in which two strands of the peptide backbone lie next to each other with hydrogen bonds formed between the two strands. The peptide strands may be oriented in the same N→C direction (parallel) or in opposite N→C directions (antiparallel). Alternate side chains in a sheet structure point toward opposite sides of the sheet.

Saccharides. Sugars.

Salt bridge. Interaction between two closely placed groups that have opposite charges.

SAM. S-adenosylmethionine. A major donor of one-carbon fragments at the methyl ($-CH_3$) oxidation state. Formed from methyl-THF and homocysteine by a vitamin B_{12}–dependent reaction.

Saturation kinetics. Michaelis-Menten. An enzyme reaction whose velocity can be described by a rectangular hyperbola, $v = V_{max}[S]/(K_m + [S])$. At high concentrations of substrate, the active site becomes completely filled or saturated with substrate, and the velocity approaches V_{max}.

Screen. A method used to tell if the DNA you want is present in a given bacterial colony. A screen doesn't select the recombinants directly by killing the cells that do or don't have your DNA; it just detects them. A screen could be a DNA hybridization to detect the presence of the DNA directly, an immunoprecipitation to detect a protein made from the DNA, or an assay for an activity of the protein.

SDS. Sodium dodecyl sulfate. A detergent used to denature proteins.

Second-order. $A + B \rightarrow C$. $v = k$ [A][B] (k in M^{-1}time^{-1}). Reactions between two molecules will go faster if you increase the concentration of either substrate.

Secondary structure. α helix, β sheet, β turn. Regular, repeating arrangements of portions of the amide backbone of a protein that allow hydrogen bonding between the carbonyl oxygens and the hydrogens on the amide nitrogens. In the folded protein, these hydrogen bonds replace hydrogen bonds that were made to water.

Selection. A way to selectively kill cells based on whether or not they contain a specific piece of DNA that encodes a protein of essential function under the selection conditions. A positive selection selects cells with the DNA of interest (they survive). A negative selection selects cells without the DNA of interest.

Shuttle vector. A DNA plasmid that contains sequences necessary for the replication of the DNA in either yeast or *E. coli*. Shuttle vectors can be used to move DNA between the two organisms.

Sialic acid. *N*-acetylneuraminic acid. $^-O_2CC(\!\!=\!\!O)CH_2CH(OH)$-CH(NH–Ac)CH(OH)CH(OH)CH(OH)CH$_2$OH. A sugar attached to extracellular proteins that is used to regulate their half-lives in the circulation (among other things). When the protein has lost its sialic acid residues because of hydrolysis in the circulation, it is cleared from the body by binding to specific receptors that recognize proteins that have lost their sialic acid.

Solvated. Surrounded by and interacting with solvent molecules. Water interacts with (solvates) ions, hydrogen-bond acceptors, and hydrogen-bond donors because of its high dielectric constant and its ability to both accept and donate multiple hydrogen bonds.

Southern. DNA–DNA blot. DNA fragments from an agarose or polyacrylamide gel are transferred to a piece of nitrocelluose or other treated paper. The "blot" is then "probed" by denaturing the DNA and annealing in the presence of a radiolabeled oligonucleotide probe. After autoradiography (placing the radioactive blot next to a piece of photographic film), bands appear where there are DNA fragments on the paper that contain sequences complementary to the probe sequence.

Specific activity. The amount of enzyme activity per milligram of protein, usually in units of micromoles of product formed per minute per milligram of protein, or units/mg.

Splicing. Removing the intervening sequences (introns) from an RNA molecule and joining the protein-coding sequences (exons).

SRP. Signal recognition particle. A protein that recognizes an NH_2-terminal protein secretion signal sequence as the protein is made on the ribosome. The SRP binds the signal sequence and inhibits translation until the ribosome attaches to the membrane of the endoplasmic reticulum.

Standard state. When equilibrium constants have units of concentration (or concentration^{-1}), the units tend to disappear when you take the logarithm to calculate the free energy. They don't really go away; you just stop writing them down. Since the log of 1000 mM is 3 and the log of 1 M is 0, the units you choose to express the equilibrium constant do affect the ΔG you calculate. When you pick your units, you have picked your "standard state." If the units are M or M^{-1}, you're dealing with a 1 M standard state.

Starvation. Lack of food intake. After about 24 hours without food, glycogen stores are depleted and the body must turn to other sources to find glucose equivalents—namely, protein degradation. Starvation is associated with glycogen depletion, increased gluconeogenesis in the liver, increased triglyceride breakdown and β oxidation, increased protein degradation, and the production of ketone bodies.

Steady-state approximation. A trick used to simplify the derivation of the rate equations for enzyme reactions. One assumes that after a short time (the pre–steady state), the rates of formation and breakdown of the various intermediates involved in the reaction become constant and equal.

Stereoisomers. Isomers that differ in the configuration around one or more asymmetric centers. If two stereoisomers have a different configuration at every center, they are enantiomers. If they differ at only one center, they're epimers. And if they differ at least two centers, they're diastereomers.

TCA. Tricarboxylic acid cycle. Krebs cycle. Citric acid cycle. 12 ATPs from acetyl-CoA → 3NADH + 2FADH$_2$ + GTP. Takes in acetyl-CoA and burns it to CO_2. The activity of the cycle is regulated by the supply of acetyl-CoA and NAD.

Telomere. Specialized structure at the ends of chromosomes that allows replication of the extreme 5' ends of the DNA without loss of genetic information. Exactly how it works is a mystery.

Tetrahedral. Arranged at the corners of a regular tetrahedron. When carbon forms bonds to four other atoms, its geometry is tetrahedral.

THF. Tetrahydrofolate. A reduced form of folic acid involved intimately in one-carbon transfer reactions. Methyl tetrahydrofolate is more reduced than methylene tetrahydrofolate, which is more reduced than methenyltetrahydrofolate.

Thiamine pyrophosphate. A cofactor involved in the pyruvate and α-ketoglutarate dehydrogenase reactions (TCA cycle) and in the transketolase reaction of the hexose monophosphate pathway.

Thiol. R–SH. A sulfhydryl group. Contained in cysteine and coenzyme A.

trans. On opposite sides. For a carbon–carbon double bond, *trans* designates a geometry in which the substituents are arranged on the opposite sides of the double bond. For peptide bonds, *trans* indicates a configuration in which the two α carbons are arranged on the opposite sides of a peptide bond. Also used to describe the effect of a product from one gene on the transcription of another gene.

Transcription. The synthesis of RNA using a DNA template.

Transfection. Putting some DNA into some organism. In bacteria this implies that you're using a viral vector.

Transferase. An enzyme that transfers something from one substrate to another.

Transformation. Putting some DNA into some organism. For eukaryotic cells this term is used to denote the process in which a normal cell is converted to a malignant cell.

Transition-state analogs. Enzyme inhibitors that rely on the idea that the active site of an enzyme is more complementary to the transition state for a chemical reaction than to the substrate. Transition-state analogs are molecules that mimic the structure of the transition state for a specific enzyme.

Transition state. During a chemical reaction, an arrangement of atoms in which bonds are partially formed or broken. It is the least stable arrangement of atoms on the pathway between reactants and products. The bigger the energy difference between the reactants and the transition state (activation energy), the slower the reaction.

Translation. The synthesis of protein directed by an mRNA. Translation occurs on ribosomes.

Triglyceride. A glycerol molecule with three fatty acid chains attached through ester linkages.

β turn. Secondary structure. An element of secondary structure in which the backbone reverses direction.

Turnover number. k_{cat} (pronounced "kay kat"). Another way of expressing V_{max}. The k_{cat} is a specific activity in which the amount of enzyme is expressed in micromoles rather than milligrams [μmol product/(min · μmol enzyme)]. The k_{cat} is the first-order rate constant for the conversion of the enzyme substrate complex into free enzyme and product.

Uncompetitive. A type of enzyme inhibition in which the inhibitor binding does not totally prevent the binding of the substrate. A pattern of parallel lines in a double-reciprocal plot of $1/v$ against $1/[S]$.

Uncouplers. Chemicals that allow protons back into the mitochondria without making any ATP. Uncouplers do not prevent the oxidation of substrates by the mitochondria, but they do prevent the formation of ATP.

Unit. The amount of enzyme required to catalyze the conversion of 1 μmol of substrate to product in 1 min. A unit of enzyme is an amount; a unit/mL is a concentration. The specific activity is units/mg; and the turnover number is units/μmol enzyme.

Urea. NH_2-C(=O)-NH_2. The major form of nitrogen excretion in mammals. Urea is also a protein denaturant that at high concentration causes proteins to lose their structure and unfold.

van der Waals interaction. A force that stabilizes protein structure. These interactions involve attractions between atoms when they come very close to each other.

Vectors. Something that can be used to introduce recombinant DNA into a host organism. An *insert* is the piece of DNA that has been placed into the vector. The most common vectors are bacterial plasmids and viruses. Vectors have been developed for specific purposes. There are cloning vectors, sequencing vectors, mutagenesis vectors, and expression vectors, each designed for a specific purpose and each with a name, like PBR322, λ-gT11, M13, etcetera. There is no logical way (and no logical reason) for remembering the specific names.

Velocity. Rate, v, activity, $d[P]/dt$, $-d[S]/dt$. How fast an enzyme converts substrate to product. The amount of substrate consumed or product formed per unit time. Units of μmol/min = units.

Viral vectors. Viruses that can be tricked into incorporating a particular gene into their genome. Bacteriophages are viruses specific for bacteria. There are also viral vectors that can be used specifically for eukaryotic cells.

V_{max}. The velocity that an enzyme-catalyzed reaction approaches at very high substrate concentration.

Western. Protein–antibody blot. After an SDS gel is run to separate proteins based on size, the proteins are transferred electrophoretically to nitrocellulose paper. The paper is then treated with a labeled antibody that specifically recognizes one protein. This antibody is labeled either with a radiolabel or with special enzyme labels. Usually the antibody is detected by another antibody that has label attached. Everywhere there was a protein on the original gel that was recognized by the first antibody, there will be a band observed on the blot.

Wild-type. The normally occurring version. The term may be applied to an organism, a gene, RNA, protein, or some college students.

Zero-order. Said of a reaction whose velocity does not change with changing substrate concentration. Zero-order behavior is observed when the concentration of an enzyme's substrate is much greater than K_m—the velocity no longer depends on substrate concentration.